PRACTICAL STATISTICS FOR FIELD BIOLOGY

Jim Fowler and
Louis Cohen

OPEN UNIVERSITY PRESS
Milton Keynes · Philadelphia

Open University Press
Celtic Court
22 Ballmoor
Buckingham MK18 1XW

and
1900 Frost Road, Suite 101
Bristol, PA 19007, USA

First Published 1990

British Library Cataloguing in Publication Data

Fowler, Jim
    Practical statistics for field biology.
    1. Biology. Applications of statistical mathematics
    I. Title   II. Cohen, Louis
    574'.01'5195

ISBN 0–335–09208–X
ISBN 0–335–09207–1 (pbk)

Library of Congress Catalog number is available

Typeset by Vision Typesetting, Manchester
Printed in Great Britain by Redwood Press Limited

CONTENTS

PREFACE

Not another introduction to statistics for biologists? We can forgive readers' suspicion that here is one more text concerned with gathering, processing, presenting and analysing biological data, one more, like many others, that makes no assumption of prior statistical knowledge on the part of potential users.

Practical Statistics for Field Biology is different. Its title signals our specific concern for those students of biology whose data are often non-standard and frequently 'messy'. Field biologists employ methods to accumulate data that are more diverse than those used by their laboratory colleagues; for one thing, they are much more concerned with handling count data. Yet introductory texts for biologists miss opportunities to help field workers relate to their data by failing to reflect the specific needs of this growing group of users. *Practical Statistics for Field Biology* seeks to redress the present imbalance.

It is our conviction that the best way to learn statistics is to apply them. This text is designed to encourage users to acquire 'statistical literacy' by getting a feel for the warp and the weft of their data and the scope of a range of statistical techniques that are applicable to their particular analytical needs. Having acquired 'literacy', students are then better placed both to use package programs and to avoid the pitfalls and temptations that such packages contain. Not that computer programs are necessary to carry out the statistical tests that are covered here; a scientific calculator is all that is required. Indeed, we invite readers to begin their introduction to statistics by reworking some of our computations. Don't be put off by minor differences between your results and ours; they arise out of small rounding errors during calculation.

We have avoided proliferating the number of statistical techniques that could be described both for the reasons implicit above and the limitations of space imposed upon us by our Editors. Whatever faults or omissions now remain we readily attribute to each other.

Finally, we dedicate this book to our wives, Eurgain and Joyce, small recompense, indeed, for many hours of isolation.

1 INTRODUCTION

1.1 What do we mean by statistics?

Statistics are a familiar and accepted part of the modern world, and already intrude into the life of every serious biologist. We have statistics in the form of annual reports, various censuses, distribution surveys, museum records – to name just a few. It is impossible to imagine life without some form of statistical information being readily at hand.

The word **statistics** is used in two senses. It refers to collections of quantitative information, and methods of handling that sort of data. A society's annual report, listing the number or whereabouts of interesting animal or plant sightings, is an example of the first sense in which the word is used. Statistics also refers to the drawing of inferences about large groups on the basis of observations made on smaller ones. Estimating the size of a population from a capture–recapture experiment illustrates the second sense in which the word is used.

Statistics, then, is to do with ways of organizing, summarizing and describing quantifiable data, and methods of drawing inferences and generalizing upon them.

1.2 Why is statistics necessary?

There are two reasons why some knowledge of statistics is an important part of the competence of every biologist. First, statistical literacy is necessary if biologists are to read and evaluate their journals critically and intelligently. Statements like, 'the probability that a first-year bird will be found in the North Sea is significantly greater than for an older one, $\chi^2 = 4.2$, $df = 1$, $P < 0.05$', enable the reader to decide the justification of the claims made by the particular author.

A second reason why statistical literacy is important to biologists is that if they are going to undertake an investigation on their own account and present their results in a form that will be authoritative, then a grasp of statistical principles and methods is essential. Indeed, a programme of work should be planned anticipating the statistical methods that are appropriate to the eventual analysis of the data. Attaching some statistical treatment as an afterthought to make the study seem more 'respectable' is unlikely to be convincing.

1.3   Statistics in field biology

'Laboratory' biologists may have high levels of confidence in the precision and accuracy of the measurements they make. To them, collecting meadow dwelling insects with a sweep net might appear a hilarious exercise with a ludicrously low level of reliability. Field biologists therefore require special sampling procedures and analytical methods if their assertions are to be regarded with credibility. Often data accumulated do not conform to the sort of symmetrical patterns taken for granted in the common statistical techniques; data may be 'messy', irregular or asymmetrical. Special treatments may be necessary before they can be properly evaluated.

1.4   The limitations of statistics

Statistics can help an investigator describe data, design experiments, and test hunches about relationships among things or events of personal interest. Statistics is a tool which helps acceptance or rejection of the hunches within recognized degrees of confidence. They help to answer questions like, *'If my assertion is challenged, can I offer a reasonable defence?'*, *'Am I justified in spending more time or resources in pursuing my hunch?'*, or *'Can my observations be attributable to chance variation?'*.

   It should be noted that statistics never *prove* anything. Rather they will indicate the likelihood of the results of an investigation being the product of chance.

1.5   The purpose of this text

The objectives of this text stem from the points made in Sections 1.2 and 1.3 above. First, the text aims to provide field biologists with sufficient grounding in statistical principles and methods to enable them to read and understand research reports in the journals they read. Second, the text aims to present biologists with a variety of the most appropriate statistical tests for their problems. Third, guidance is offered on ways of presenting the statistical analyses, once completed.

2 MEASUREMENT AND SAMPLING CONCEPTS

2.1 Populations, samples and observations

Biologists are familiar with the term **population** as meaning all the individuals of a species that interact with one another to maintain a homogeneous gene pool. In statistics, the term population is extended to mean any collection of individual items or **units** which are the subject of investigation. Characteristics of a population which differ from individual to individual are called **variables**. Length, mass, age, temperature, proximity to a neighbour, number of parasites, number of petals, to name but a few, are examples of biological variables to which numbers or values can be assigned. Once numbers or values have been assigned to the variables they can be measured.

Because it is rarely practicable to obtain measures of a particular variable from all the units in a population, the investigator has to collect information from a smaller group or **sub-set** which represents the group as a whole. This sub-set is called a **sample**. Each unit in the sample provides a record, such as a measurement, which is called an **observation**. The relationship between the terms we have introduced in this section is summarized below:

Observation: 132 mm
Variable: wing length
Sample unit: a starling from a communal roost
Sample: those starlings which are captured in the roost and are measured
Statistical population: all starlings in the roost which are available for capture and measurement
Biological population: the biological population may well include birds that are not available for capture (e.g. mates that are roosting elsewhere) and are therefore not part of the statistical population. Alternatively, if the roost comprises a mixture of resident birds and winter immigrants, the statistical population might include components of more than one biological population.

2.2 Counting things – the sampling unit

Field biologists often count the number of objects in a group or collection. If the number is to be meaningful, the dimensions of the collection have to be specified. A collection with specified dimensions is called a **sampling unit**; a set of sampling units comprise a sample. An observation is, of course, the number of objects in a particular sampling unit. Examples of sampling units are:

Observation	Sampling unit
Number of orchids	A quadrat of stated area
Number of crickets in a sweep net	Volume of vegetation swept (diameter of net × distance moved)
Number of nematodes in a soil core	Soil core of stated dimensions
Number of visits by bees to a flower	A specified interval of time
Number of wading birds on a shore	A specified length of coastline
Number of ectoparasites	A single host
Number of beetles in a pitfall trap	A trap of stated size

When observations are counts, the statistical population has nothing to do with the objects we are counting, even when they are organisms. The following example illustrates the point.

Observation: 23
Variable: number of cockles
Sampling unit: a square quadrat of stated area from which sand is sieved and cockles counted
Sample: the number of quadrats (sampling units) examined
Statistical population: the total number of quadrats it is *possible* to mark out in the whole of the study area. The potential number of units in the population depends on the chosen dimensions of the sampling unit.

The main difference between 'measuring' and 'counting' is that we have no control over the dimensions of a unit in a sample when we are measuring; when counting, we *are* able to choose the dimensions of the sampling unit. Remember that the content of a trap, net or quadrat is a sample if we are measuring the objects in it, but only a unit in a sample if we are counting them.

It is always worthwhile to ask the question, '*from which population are my sampling units drawn?*'. The answer may not always be as obvious as in the example of the cockles. The contents of 10 pit-fall traps set into the ground overnight constitute a sample – but from which population are these sampling units drawn? It is regarded as being the total number of traps that *could* have been set out, covering the whole of the study area. Because it is axiomatic that field biologists try not to destroy the habitat they are studying, a statistical population is sometimes notional, or **hypothetical**.

2.3 Random sampling

We say in Section 2.1 that a sample *represents* the population from which it is drawn. If the sample is to be truly representative, the units in the sample must be drawn **randomly** from the population; that is to say, in a manner that is free from **bias**. In other words, each unit in a population must have an equal chance of being drawn.

As an example of a possible source of bias, consider a biologist who wishes to measure the average mass of bank voles *Clethrionomys glareolus* inhabiting a study site. Attempts are made to catch them by setting Longworth mammal traps baited with grain. Before capture, an animal has to overcome trap shyness. It is plausible that the threshold of shyness is lower in hungry animals than in well-fed ones and that the former may have a greater chance of being drawn from the population. If hungry voles are lighter than well-fed ones, our biologist's sample may not be a fair representation of the whole population.

Statistical analysis is frequently conducted on the assumption that samples are random. If, for any reason, that assumption is false and bias is present in the sampling procedure, then the information gained from the sample may not be properly extrapolated to the population. Unfortunately, it is rarely possible to do more than guess how great bias may be. This severely reduces the confidence which can be placed in estimations based on sampling data. Since most sources of bias arise from the methodology adopted, procedures should always be fully described. When a source of bias is suspected, it should be acknowledged and taken into account in the interpretation of results. The practical aspects of obtaining random samples is a large area in itself, partly because the field techniques used by biologists are so diverse. We suggest you consult Southwood (1978) as a standard work on this subject (see Bibliography).

2.4 Random numbers

One way to avoid bias is to assign a unique number to each individual unit in a population and select units to be measured by reference to random numbers. Often this is impossible because we cannot always choose our units – we measure what we can catch, as in the example of the voles. However, it is sometimes possible – indeed essential – to obtain truly random sampling units. In the case of our cockle example in Section 2.2, the quadrats comprising the sample could be located at the intersection of grid coordinates prescribed by pairs of random numbers. Whenever there is opportunity to select 'which plots?', 'which pools?', or 'which positions?', then selection must be based on random numbers.

There are two usual ways of obtaining random numbers. First, many calculators and pocket computers have a facility for generating random numbers. These are often in the form of a fraction, e.g. 0.2771459. You may use this to provide a set of integers, 2, 7, 7, 1, . . ., or 27, 71, 45, . . .; or 277, 145, . . .; or 2.7, 7.1, . . .; and so on, keying a new number when more digits are required.

Second, use may be made of random number tables. Appendix 1 is such a table. The numbers are arranged in groups of five in rows and columns, but this arrangement is arbitrary. Starting in the top left corner you may read, 2, 3, 1, 5, 7, 5, 4, . . .; or 23, 15, 75, 48, . . .; or 231, 575, 485, . . .; or 23.1, 57.5, 48.5, 90.1, . . .; and so on, according to your needs. When you have obtained the numbers you need for the investigation in hand, mark the place with a pencil. Next time, carry on where you left off.

It is possible, by chance, that a random number will prescribe a unit that has already been drawn. In this event, ignore the number and take the next random number. The purpose is to eliminate *your* prejudice as to which units should be picked for measurement or counting. Unfortunately, **observer bias**, conscious or subconscious, is notoriously difficult to avoid when gathering data in support of a particular hunch!

2.5   Independence

Many statistical methods assume that observations in a sample are **independent**. That is to say, the value of any one observation in a sample is not inherently linked to that of another. An example should make this clear. A biologist wishes to compare the average spikelet length of rough meadow grass growing in one field with that growing in another. One hundred flowering heads are obtained randomly from the first field, a spikelet is removed from each and measured. In the second field, the plant is harder to find and only 80 flower heads are collected, a spikelet being removed from each and measured. If the biologist now tries to 'make up the number' by removing a further 20 spikelets from one plant, these observations are not independent of each other even if the plant itself is randomly selected. A genetic peculiarity in the plant that affects the size of one spikelet is likely to affect them all. This may distort the sample (see also Section 13.4).

2.6   Statistics and parameters

The measures which describe a variable of a sample are called **statistics**. It is from the sample statistics that the **parameters** of a population are estimated. Thus, the average mass of a random sample of voles is the statistic which is used to estimate the average mass parameter of the population. The average number of cockles in a random sample of quadrats estimates the average number of cockles per quadrat in the whole population of quadrats.

Hypothetical populations have hypothetical parameters. The average number of beetles in 10 randomly placed pit-fall traps estimates the average number of beetles per trap if the whole habitat had been covered by traps, in which case there are no beetles left to count! Samples from hypothetical populations are generally used for **comparative** purposes, for example to compare one woodland type with another.

In estimating a population parameter from a sample statistic, the number of units in a sample can be critical. Some statistical methods depend on a minimum number of sampling units and, where this is the case, it should be borne in mind before commencing fieldwork. Whilst it is true that larger samples will invariably result in greater statistical confidence, there is nevertheless a 'diminishing returns' effect. In many cases the time, effort and expense involved in collecting very large samples might be better spent in extending the study in other directions. We offer guidance as to what constitutes a suitable sample size for each statistical test as it is described.

2.7 Descriptive and inferential statistics

Descriptive statistics are used to organize, summarize and describe measures of a sample. No predictions or inferences are made regarding population parameters. **Inferential** (or **deductive**) **statistics**, on the other hand, are used to infer or predict population parameters from sample measures. This is done by a process of inductive reasoning based on the mathematical theory of **probability**. Fortunately, only a very minimal knowledge of mathematical theory of probability is needed in order to apply the rules of the statistical methods, and the little that is needed will be explained. However, no one can predict exactly a population parameter from a sample statistic, but only indicate with a stated degree of confidence within what range it lies. The degree of confidence depends on the sample selection procedures and the statistical techniques used.

2.8 Parametric and non-parametric statistics

Statistical methods commonly used by biologists fall into one of two classes – **parametric** and **non-parametric**. Parametric methods are the oldest, and although most often used by statisticians, may not always be the most appropriate for analysing biological data. Parametric methods make strict assumptions which may not always hold true.

More recently, non-parametric methods have been devised which are not based upon stringent assumptions. These are frequently more suitable for processing biological data. Moreover they are generally simpler to use since they avoid the laborious and repetitive calculations involved in some of the parametric methods. The circumstances under which a particular method should be used will be described as it arises. A summary showing which methods should be applied in particular circumstances is provided in Section 12.8.

3 PROCESSING DATA

3.1 Scales of measurement

Variables measured by biologists can be either **discontinuous** or **continuous**. Values of discontinuous variables assume integral whole numbers and are usually **counts** of things (frequencies). On the other hand, values of continuous variables may, in principle, fall at any point along an uninterrupted scale, and are usually measurements (length, mass, temperature, etc.). Measurement values may sometimes appear to be integral whole numbers if the recorder elects to measure to the nearest whole unit; this does not, however, obviate the fact that there can be intermediate values. The distinction between 'count data' and 'measurement data' is an important one which will be referred to frequently.

Generally, four levels of measurement are recognized. They are referred to as **nominal**, **ordinal**, **interval** and **ratio** scales. Each level has its own rules and restrictions; moreover each level is hierarchical in that it incorporates the properties of the scale below it.

3.2 The nominal scale

The most elementary scale of measurement is one which does no more than identify categories into which individuals may be classified. The categories have to be mutually exclusive, i.e. it should not be possible to place an individual in more than one category. The nominal level of measurement is often used by biologists. For example, species, sex, colour and habitat type are all nominal categories into which count data can be assigned.

The name of a category can of course be substituted by a number – but it will be a mere label and have no numerical meaning. Thus, if blue tits are coded 1, coal tits 2, great tits 3, willow tits 4 and marsh tits 5 they can then be listed, 1,2,3,4,5 but the sequence has no more mathematical significance than if they had been listed 4,2,1,5,3. They are still nominal categories.

3.3 The ordinal scale

The ordinal scale incorporates the classifying and labelling function of the nominal scale, but in addition brings to it a sense of order. Ordinal numbers are used to indicate **rank order**, but nothing more. The ordinal scale is used to arrange (or rank) individuals into a sequence ranging from the highest to the lowest, according to the variable being measured. Ordinal numbers assigned to such a sequence may not indicate absolute quantities, nor can it be assumed that intervals between adjacent numbers on the scale are equal.

An example of an ordinal scale is the ***DAFOR* scale** used to record the abundance of different plant species in a quadrat:

	score
Dominant	5
Abundant	4
Frequent	3
Occasional	2
Rare	1

In this scale there is no simple relationship between the numerical values of the abundance scale. 'Abundant' does not mean twice 'occasional', but it will always be ranked above 'frequent'.

3.4   The interval scale

As the term **interval** implies, in addition to rank ordering data, the interval scale allows the recognition of *precisely how far apart* are the units on the scale. Interval scales permit certain mathematical procedures untenable at the nominal and ordinal levels of measurement. Because it can be concluded that the difference between the values of, say, the 8th and 9th points on the scale is the same as that between the 2nd and 3rd, it follows that the intervals can be added or subtracted. But because a characteristic of interval scales is that they have *no absolute zero point* it is *not* possible to say that the 9th value is three times that of the 3rd. To illustrate this, date is a very widely used interval scale. If the first-arrival dates of four species of warbler are, respectively, the 1st, 5th, 10th and 15th May, the interval between each point on the scale (1 day) is equal, and the fourth species took 10 days longer to arrive than the second. It did not take three times as long, however, any more than it took 15 times longer to arrive than the first species! Another interval scale is temperature: 10°C is not twice as hot as 5°C because the zero on the scale in question (Celsius) is not absolute.

3.5   The ratio scale

The highest level of measurement, which incorporates the properties of the interval, ordinal and nominal levels, is the ratio scale. A ratio scale includes an absolute zero, it gives a rank ordering and it can simply be used for labelling purposes. Because there is an absolute zero, all of the mathematical procedures of addition, subtraction, multiplication and division are possible. Measurements of length and mass fall on ratio scales. Thus, a length of 150 mm *is* three times as long as one of 50 mm.

The mathematical properties of interval and ratio scales are similar and as no statistical procedure described here will distinguish between them, we shall refer to them both as 'interval' scales.

3.6   Conversion of interval observations to an ordinal scale

Usually, observations made on interval scales allow the execution of more sensitive statistical analyses. Sometimes, however, interval data are not suitable for certain methods. Perhaps, because there are too few observations, we are forced to downgrade them to an ordinal rank scale for use in non-parametric methods. The following measurements (mm) are ranked in increasing size in the top line. Their rank (ordinal) scores are underneath:

Length (mm):	31.0	31.4	32.3	33.1	33.5	34.9	35.0	36.6	37.2	38.0
Rank score:	1	2	3	4	5	6	7	8	9	10

If large numbers of observations of a variable are collected, it is almost inevitable that some of the observations will be equal in value. Their ranks will also be tied and these have to be dealt with correctly. Since some statistical tests which we describe later depend on the ranking of observations, we take the opportunity now of dealing with the problem of **tied observations**.

Where tied observations occur, each of them is assigned the average of the ranks that would have been given if there had been no ties. To illustrate this, a set of measurements rounded to the nearest whole number is given below. For convenience they are presented in ascending order and adjacent tied scores are underlined:

25   26   27   27   28   29   30   30   30   31   32   33   33   33   33   34
35   36   36   36   36   36   37

If we try to rank these, the single extreme values of 25 and 37 will clearly be ranked 1 and 23, respectively. The two values of 27 together occupy the ranks of 3 and 4; they are each assigned the average rank of $3\frac{1}{2}$. The three values of 30 occupy the ranks 7, 8 and 9. They have an average rank of 8. In similar manner, the four values of 33 are each assigned the rank $13\frac{1}{2}$ and the 5 of 36 the rank of 20. The set of data is rewritten below, with the correct ranks assigned:

Observation:	25	26	27	27	28	29	30	30	30	31	32	33	33
Rank:	1	2	$3\frac{1}{2}$	$3\frac{1}{2}$	5	6	8	8	8	10	11	$13\frac{1}{2}$	$13\frac{1}{2}$

Observation:	33	33	34	35	36	36	36	36	36	37
Rank:	$13\frac{1}{2}$	$13\frac{1}{2}$	16	17	20	20	20	20	20	23

3.7   Derived variables

Sometimes observations are processed in order to generate a **derived number**. Examples of derived variables are *ratios, proportions, percentages,* and *rates.*

A **ratio** is the simple relationship between two numbers. If the width of the head capsule of a feather louse is 1.24 mm and the length is 2.15 mm, the width:length ratio is 1.24:2.15. Alternatively, the length:width ratio is 2.15:1.24. One value in a pair is sometimes converted to unity by division. A sample which contains 18 males and 25 females has a male:female ratio of 18:25 or 1:25/18, that is, 1:1.39. A ratio may be expressed as a fraction. In the last example, the ratio of males to females is $18/25 = 1/1.39$. When the fraction is reduced to a decimal number, it is often called a *coefficient*; thus, $1/1.39 = 0.72$.

A **proportion** is the ratio of a *part* to the *whole*. If the total body length (head + thorax + abdomen) of the louse in the example above is 6.3 mm, the proportion of the body length accounted for by the head is $2.15:6.3 = 0.34$. If a proportion is based on counts of things it may be referred to as a *proportional frequency*, that is, the ratio of the number of individuals in a particular category to the total number in all categories.

Example 3.1

The number of four species of woodlice in a pit-fall trap are: *Oniscus* 12; *Porcellio* 8; *Philoscia* 5; *Armadilidium* 2. What is the proportion of each species in the sample?

The total number N of woodlice is $12 + 8 + 5 + 2 = 27$.
The proportion is given by:

$$p_i = \frac{n_i}{N}$$

where p_i is the proportion of a particular category, n_i is the number of individuals in a particular category and N is the total number in all categories. Therefore:

Species	n_i	n_i/N	p_i	Percentage
Oniscus	12	12/27	0.444	44.4
Porcellio	8	8/27	0.296	29.6
Philoscia	5	5/27	0.185	18.5
Armadilidium	2	2/27	0.074	7.4
	$N = 27$		$\Sigma p_i = 1$	

Notice that the sum of the individual proportions equals 1.

When a proportion is multiplied by 100 it is called a **percentage**. The percentage of each species of woodlice in the sample described in Example 3.1 is included in the table above.

A **rate** is the ratio of an observation to a period of time. Rates are useful for expressing such variables as growth, population change, and movement.

Example 3.2

A shoot grows 12 cm in 4 days.
Ratio $= 12:4 = 3:1$.   Rate $= 3$ cm/day

A pigeon flies 1728 km in 24 h.
Ratio $= 1728:24 = 72:1$.   Rate $= 72$ km/h.

Other derived variables we refer to in this text are the **Lincoln index** (Section 11.9) and a **diversity index** (Section 11.10). Statistical techniques may be performed upon derived variables; sometimes the data first have to be converted or *transformed* (see, for example, Section 10.5).

3.8   The precision of observations

When an observation is of a discrete variable, that is a **count**, we are usually sure of its precision. A nest may have *exactly* four eggs in it. A measurement, on the other hand, is never exact; it is only precise to within certain limits.

If the diameter of a pond is measured with a tape marked in 1-metre intervals, measurements are precise to the nearest whole metre. An observation of 10 m can be recorded as 10 ± 0.5 m. This implies that all distances between the limits of 9.5 m and 10.5 m are recorded as 10 m, as shown in Fig. 3.1.

We could use a more finely graduated scale, for example a tape marked in 10 cm (0.1 m) intervals. Each observation is then precise to the nearest 0.1 m and an observation of 10.6 m is written as 10.6 ± 0.05 m; that is, all distances between 10.55 m and 10.65 m are recorded as 10.6 m.

We could continue increasing the precision of measurements by using a metre rule graduated in millimetres, Vernier callipers capable of recording 0.01 mm, or a microscope eye-piece graticule down to 0.001 mm. In each case, the measurement is precise to *plus* or *minus half the interval spanned by the last measured digit*; in the case of the graticule measurement this is ± 0.0005 mm. An observation of 1.364 mm is within an interval spanning 1.3635 mm to 1.3645 mm.

Note the distinction between *precision* and *accuracy*. An expensive spring balance might be precise, weighing to 0.1 g, but if it is badly adjusted it will not be accurate. A broken clock is accurate twice a day!

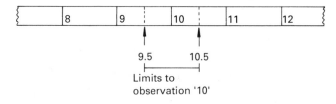

Fig. 3.1   The limits of an observation.

3.9 How precise should we be?

Since it is clearly possible to choose (within reason) the degree of precision of a measurement, the question arises, 'how precisely should we make our measurements?'. We should not choose, for example, to measure a transect across a salt-marsh in millimetres.

The authors Sokal and Rohlf (1981) suggest that the degree of precision should be such that there are between 30 and 300 unit steps between the largest and smallest observations. For example, ornithologists sometimes measure the wing length of birds to the nearest whole millimetre. If, in a sample of blue tits, the largest observation is 67 mm and the smallest is 59 mm then there are $(67 - 59) = 8$ unit steps, well below the suggested lower limit of 30. An error of 1 unit in 8 represents an unacceptable 12.5%. Measurement to a tenth of a millimetre would give about 80 steps. We do not know exactly how many steps because the largest length could have been, say, 67.4 mm and the smallest 58.8, a difference of 86 unit steps. This is more satisfactory – an error of 1 unit in 80 is an acceptable 1.25%.

3.10 The frequency table

Collected data should be organized and summarized in a form that allows further interpretation and analysis. In Table 3.1 are the lengths (to the nearest whole millimetre) of 100 shoots grown from seeds that were planted at the same time.

The measurements are presented in the order that the shoots were measured and are therefore *ungrouped*. A quick scan of Table 3.1 reveals that particular values are repeated a number of times: there are, for example, five values of 74 mm in the top row alone. The value of *74 mm* is called a **frequency class** and, rather than record all 100 values individually (as in the table), it is more economical of space and more revealing to *group* the data into all the frequency classes. We should remember that each frequency class is a **class interval** with

Table 3.1 Lengths of 100 shoots (mm)

76	73	75	73	74	74	74	74	74	77
74	72	75	76	73	71	73	80	75	75
68	72	78	74	75	74	69	77	77	72
72	76	76	77	70	77	72	74	77	76
78	72	70	74	76	72	73	71	74	74
75	79	75	74	75	74	71	73	75	73
75	70	73	75	70	72	72	71	76	73
74	76	74	75	74	76	75	75	73	73
78	74	73	75	74	73	72	76	73	76
74	71	72	71	79	78	69	77	73	71

Table 3.2   Grouped lengths of 100 shoots

Implied class interval	Frequency class x (mm)	Tallies	Frequency f
67.5–68.5	68	1	1
68.5–69.5	69	11	2
69.5–70.5	70	1111	4
70.5–71.5	71	1111111	7
71.5–72.5	72	11111111111	11
72.5–73.5	73	111111111111111	15
73.5–74.5	74	11111111111111111111	20
74.5–75.5	75	111111111111111	15
75.5–76.5	76	11111111111	11
76.5–77.5	77	1111111	7
77.5–78.5	78	1111	4
78.5–79.5	79	11	2
79.5–80.5	80	1	1

$$n = \Sigma f = 100$$

implied limits of ± 0.5 mm. Thus, all lengths between 73.5 mm and 74.5 mm are placed in the '74 mm' class. The grouped observations are shown in Table 3.2.

The data in Table 3.2 are grouped into columns: implied class interval, frequency class x and frequency f. The *tallies* are presented in this example to give a visual appreciation of how the frequencies are distributed between the frequency classes. The two columns x and f represent a **frequency table**. Although this particular table has been constructed for interval measurements, frequency tables can also be constructed for count data and for nominal and ordinal scales. The manner in which frequencies are distributed between the frequency classes is described as a **frequency distribution**.

Readers will note that there are only 12 unit steps between the largest and smallest observations in the table above and an increase in precision is highly desirable. The purpose of the hypothetical data is simply to illustrate the construction of a frequency table.

3.11   Aggregating frequency classes

When the spread of observations is large and the number of observations relatively few, then a frequency distribution appears drawn out and disjointed if every step interval corresponds to a frequency class. In such cases it is advisable to aggregate, or group adjacent classes to smooth out the distribution. The following example shows how to do this.

Example 3.3

A sample of 50 migratory salmon from a single young cohort is weighed. Salmon masses (g) are presented in Table 3.3. We note first that there are $(200-156)=44$ unit steps between the largest and smallest observation; the data are sufficiently precise. Second, each observation is measured to the nearest whole gram and is therefore precise to within ± 0.5 g. The first observation (162) thus falls within an implied interval of 161.5 g to 162.5 g.

Table 3.3 Masses of 50 salmon

162	188	173	168	174	183	167	186	177	187
170	174	164	174	159	177	173	163	180	196
171	156	184	179	190	181	166	181	182	176
169	172	174	162	175	192	178	177	200	191
188	168	165	179	193	175	160	180	187	176

The data are summarized and grouped into the frequency table in Table 3.4.

Table 3.4 A grouped frequency table

Implied limits of each class interval	*Class mark (mid-point of class)*	*Tallies*	*Frequency f*
155.5–158.5	157	1	1
158.5–161.5	160	11	2
161.5–164.5	163	1111	4
164.5–167.5	166	111	3
167.5–170.5	169	1111	4
170.5–173.5	172	1111	4
173.5–176.5	175	11111111	8
176.5–179.5	178	111111	6
179.5–182.5	181	11111	5
182.5–185.5	184	11	2
185.5–188.5	187	11111	5
188.5–191.5	190	11	2
191.5–194.5	193	11	2
194.5–197.5	196	1	1
197.5–200.5	199	1	1
			$n=\Sigma f=50$

The steps involved in the construction of Table 3.4 are listed below. They enable the construction of frequency tables from any set of measurement data with a degree of class aggregation that suits specific needs.

1. Determine the *range* of scores: the highest observation minus the lowest observation plus 1. (One is added to take into account the implied limits of the numbers.)

$$\text{Range} = \text{highest observation } (200) - \text{lowest observation } (156) + 1$$
$$= (200 - 156) + 1$$
$$= 45$$

Decide how many categories (class intervals) are required. Normally the number of class intervals is not less than 10 and not more than 20. In this case we have selected 15 as a convenient number of classes.

2. Divide the range by the number of class intervals required. This gives the number of unit steps which are aggregated to make up an interval class.

$$\text{Number of unit steps per class interval} = \frac{45}{15} = 3$$

If the calculated number of unit steps per class interval is a fraction, then round to the nearest whole number.

3. Construct the class interval column starting at the top with the lower limit of the smallest observation (155.5). Add the class interval size (3) to this lower limit. The range of the lowest class interval becomes 155.5 to 158.5. The lower limit of the next class becomes 158.5 to which 3 is added to give the class interval range 158.5 to 161.5. This procedure is repeated, moving down the column until the class interval column includes an interval into which the largest observation (200) can be placed, namely, 197.5 to 200.5.

4. Insert in the next column the mid-point of the class. This labels the class, and is called the **class mark**. It is obtained by adding half the class interval size to the lower limit of the class; thus, $155.5 + 3/2 = 157$.

5. Insert in the next column provided a tally for each individual observation in the raw data table. For example, for the observation 162, a tally is inserted to show that it falls into the class interval range 161.5 to 164.5.

6. Total up the tallies within each class interval and place in the frequency column in line with the appropriate class.

7. Total the frequency column (*n*). This serves as a useful check that all data have been included in the table.

3.12   Frequency distribution of count observations

In principle, the construction of a frequency distribution of count observations is similar to that of measurement observations. If the observations in Section 3.10 were counts, for example, the number of daisy flowers in 100 quadrats on a lawn, the frequency distribution would be exactly the same. However, because each observation, and hence, frequency class has an *exact* value, the column of *implied class interval* is redundant. There is no implied upper and lower limit to a count of 74 daisies.

A minor problem may arise if frequency classes are aggregated. Imagine that the observations in Example 3.3 are counts, for example, the number of protozoa counted in 50 sampling units of pond water. The construction of a frequency table is undertaken exactly as described, except that the column of implied limits is again redundant. The class mark (the mid-point of each class interval) is still a useful number, however.

Suppose that we had elected to aggregate the unit steps into classes of four instead of three. In Step 4 of our instructions, the range of the lowest class interval would be 155.5 to 159.5. To find the class mark we would add half of the class interval range to the lower limit, that is, $155.5 + 4/2 = 157.5$. A class mark that is a *fraction* is not sensible in a frequency distribution of counts; a frequency class labelled '157.5 protozoa' has little meaning. To construct a frequency distribution of count observations in which classes are aggregated, make sure that the number of units steps in each class interval is an *odd number*. The class mark is then a whole number.

3.13 Dispersion

The word **distribution** calls for as much care in its usage as the word **population** when biologists set out to describe and to analyse their data by statistical methods.

As we have seen, distribution has a special meaning in statistics to do with how observations of a variable are spread over the range of measurement. Biologists commonly use the word distribution to mean how organisms are scattered about in the environment. Thus, 'the pipistrelle bat has a wide distribution' or 'gull nests are distributed evenly through the colony'.

To avoid risk of confusion, we adopt the word **dispersion** to indicate how or where objects are placed in the environment.

3.14 Bivariate data

Biologists frequently obtain more than one observation from a unit in a sample. A vole may be weighed and measured; a quadrat may provide a count of plants and a soil pH value; a whiting may furnish the length of the fish and the length of an otolith dissected from it.

A set of observations of *two* variables from each item or unit in a sample is called **bivariate data**. There are statistical methods available for analysing bivariate data.

4 PRESENTING DATA

4.1 Introduction

One drawback in presenting data in the form of a frequency distribution table is that the information contained there does not become immediately apparent unless the table is studied in detail. To simplify the interpretation of the information, and to pin-point patterns and trends, the data are often processed further and transformed into a visual presentation. The most common methods of presenting data are based upon graphical techniques. In this section we describe methods suitable for presenting biological data.

4.2 Dot diagram or line plot

The dot diagram is a method of presenting data which gives a rough but rapid visual appreciation of the way in which the data are distributed. It consists of a horizontal line marked out with divisions of the scale on which the variable is being measured. A dot representing each observation is placed at the appropriate point on the scale. If certain observations are repeated, the dots are simply stacked on top of each other. Figure 4.1 shows two dot-diagrams, each involving 16 sampling units:

(a) Lengths of 16 leaves measured to the nearest whole millimetre: observations are scattered about 63 mm.
(b) Numbers of orchids in 16 randomly placed quadrats: 7 quadrats contain no orchids.

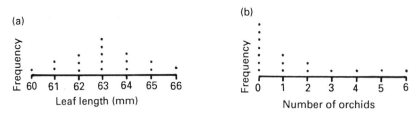

Fig. 4.1 Dot diagrams showing frequency distributions of (a) 16 leaf lengths; (b) numbers of orchids in 16 quadrats.

Although dot diagrams are invaluable for the preliminary analysis of data, they are seldom used as the form of final presentation. (See also the dot diagram in Section 5.3.)

4.3   Bar graph

Portraying information by means of a bar graph is particularly useful when dealing with data gathered from discrete variables that are measured on a nominal scale. A bar graph uses lines (i.e. bars) to represent discrete categories of data, the length of the line being proportional to the frequencies within that category.

Suppose 31 nest boxes are placed in a wood, and 15 become occupied by blue tits, 10 by great tits, 4 by tree sparrows and 2 by nuthatches. A frequency table may be constructed:

	f
blue tit	15
great tit	10
tree sparrow	4
nuthatch	2
	$n = 31$

Using these data, a bar graph may be constructed, as shown in Fig. 4.2.

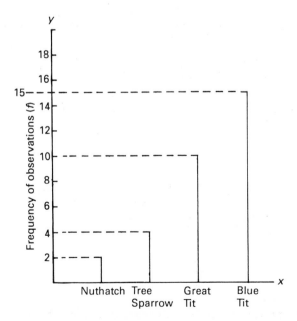

Fig. 4.2   Bar graph showing nest box occupancy.

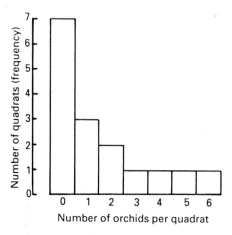

Fig. 4.3   Frequency distribution of orchid counts.

In its final form, the horizontal dashed lines are omitted; they are included here to show that the height of the bar corresponds to the respective frequency.

When observations are counts of things the bar graph is a useful way to present a frequency distribution. Illustrators often replace each bar with a vertical rectangle, or block, whose adjacent sides are touching. The frequency distribution of orchid counts shown as a dot diagram in Fig. 4.1 is shown as a bar graph in Fig. 4.3, where the height of each block is still proportional to the frequency in each category because the width of each block is equal. When presented in this form the diagram is usually referred to as a **histogram**. Histograms are especially useful for presenting frequency distributions of obervations measured on continuous variables, as we show in Section 4.4.

4.4   Histogram

The histogram is especially useful for presenting distributions of observations of **continuous variables**. In a histogram the *area* of each block is proportional to the frequency. The area of a single histogram block is found by multiplying the width of the block (the class interval) by the height (frequency).

Example 4.1

150 dead fish are recovered from a stream following a pollution incident, and are measured to the nearest whole millimetre. Measurements obtained are expressed in the form of a frequency table:

Length of fish (mm)	Number of fish (frequency)
100–109	7
110–119	16
120–129	19
130–139	31
140–149	41
150–159	23
160–169	10
170–179	3

The frequency distribution is presented in a histogram in Fig. 4.4. Because the width of each block (class interval) is the same, 10 mm, the height of each block is proportional to the frequency. In Fig. 4.4 the area of the first block is $10 \times 7 = 70$ units, compared with the fifth block which is $10 \times 41 = 410$ units. The area represented by the last three blocks is $(10 \times 23) + (10 \times 10) + (10 \times 3) = 360$ units.

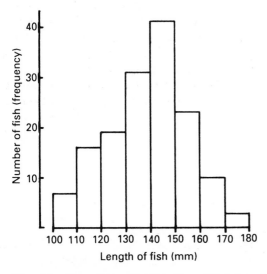

Fig. 4.4   Frequency distribution of fish lengths.

Sometimes it is convenient to *aggregate* frequency classes, especially those in the 'tails' of the distribution. The distribution of Example 4.1 is re-written in the table on the following page with the last three classes aggregated:

Length of fish (mm)	Number of fish (frequency)
100–109	7
110–119	16
120–129	19
130–139	31
140–149	41
150–179	36

In the last category, the 36 fish are aggregated into a class spanning 30 mm. The *area* of the block which will represent them in a histogram is still 360 units, as above. However, since the width of the block is 30 units, the height of the block is 360/30 = 12 units. This is, of course, the average height of the three blocks before they were aggregated. The resulting histogram is shown in Fig. 4.5.

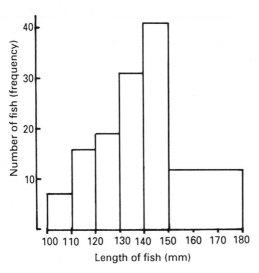

Fig. 4.5   Frequency distribution of fish lengths with aggregated tail.

4.5   Frequency polygon and frequency curve

If the mid-point of the top of each block in a histogram is joined by a straight line, a **frequency polygon** is produced. Figure 4.4 is reproduced with a frequency polygon superimposed in Fig. 4.6. When the number of observations of a continuous variable is large and the unit increments are small, the 'steps' in the histogram tend towards a smooth, continuous curve, called a **frequency curve**. A frequency curve is superimposed (in Fig. 4.7), upon the distribution of 100 shoot lengths given in Section 3.10.

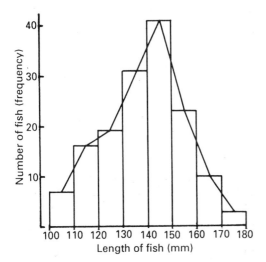

Fig. 4.6 Frequency polygon of fish lengths.

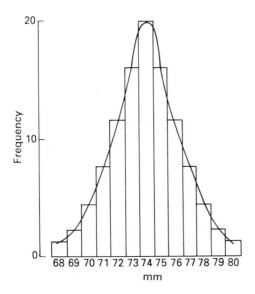

Fig. 4.7 Frequency curve of shoot length measurements.

4.6 Scattergram

When *pairs* of observations of two variables are obtained from each unit in a sample (that is, the data are bivariate), a scattergram is used to display the data. In Fig. 4.8 each point in the scattergram represents an individual animal for

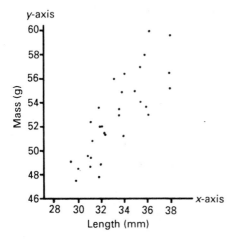

Fig. 4.8   Scattergram of mass and length observations of 32 animals.

which mass *and* length are indicated. In two-dimensional presentations like this, it is conventional to refer to the vertical scale as the *y-axis* and the horizontal scale as the *x-axis*.

4.7   Circle or pie graph

The pie graph is best suited for displaying data which are *percentages* or *proportions*. If the area of a circle is regarded as 100% it can be divided into sectors (the slices of the pie) which correspond in size to each individual percentage or proportion making up the total. To work out the angle of each sector at the centre of the pie divide each percentage by 100 and multiply by 360, the number of degrees in a circle. Proportions are simply multiplied by 360.

Example 4.2

The percentage time spent on different activities ('time budget') by feral nanny and billy goats during a rut is presented below. The number of degrees corresponding to each percentage is given in brackets.

	Feeding	Lying	Standing	Walking	Social/sexual
		Activity %			
Nannies	67	24	5	2	2
	(241°)	(87°)	(18°)	(7°)	(7°)
Billies	34	41	10	5	10
	(122°)	(148°)	(36°)	(18°)	(36°)

A pie graph of these data is shown in Fig. 4.9, where the different sectors of the pie are shaded for extra visual impact.

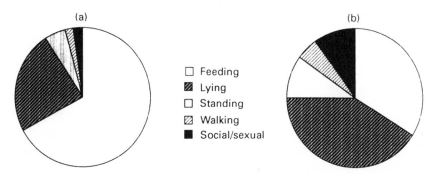

Fig. 4.9 Activity budgets of feral goats during rut. (a) nannies; (b) billies.

5 MEASURING THE AVERAGE

5.1 What is an average?

One meaning of statistics is to do with describing and summarizing quantitative data. Any description of a sample of observations must include an aspect which relates to **central tendency**. That is to say, we need to identify a single number close to the centre of the distribution of observations which represents them all. We call this number the **average**; it is often referred to as a measure of **location** because it indicates, on what might be a scale of infinite magnitude, just where a cluster of observations is located. The average is described by one of three commonly used statistics: the **mean**, the **median** and the **mode**. Each has its own application.

5.2 The mean

The mean or, more precisely, the **arithmetic mean** is the most familiar and useful measure of the average. It is calculated by dividing the sum of a set of observations by the number of observations. If it is possible to obtain an observation from every single item or sampling unit in a population, for example, the mass of every remaining living Californian condor, then the mean is symbolized by μ (mu) and is called the **population mean**. More usually we have to be content with observations from a sample, in which case the **sample mean** is symbolized by $\bar{x}$ ('x-bar'). The sample mean is a direct estimate of the population mean (thus $\bar{x}$ estimates μ). In Chapter 11 we explain how good an estimate it is likely to be. The formulae for calculating the mean are:

$$\mu = \frac{\Sigma x}{N} \quad \text{and} \quad \bar{x} = \frac{\Sigma x}{n}$$

where x is each observation, N is the number of items (observations) in a population, n is the number of observations in a sample and Σ is the 'sum of'.

Example 5.1

The maximum diameters of the pileus (cap) of five specimens of an edible fungus are recorded below.

8.5 cm   9.2 cm   7.3 cm   6.8 cm   10.1 cm

Calculate the mean diameter.

1. Obtain the sum of the observations (Σx):

$\Sigma x = 8.5 + 9.2 + 7.3 + 6.8 + 10.1 = 41.9$ cm

2. Divide Σx by n, the number of observations:

$$\bar{x} = \frac{\Sigma x}{n} = \frac{41.9}{5} = 8.38 \text{ cm}$$

Conventionally, the mean is recorded to one more decimal place than the original data if n is less than 100, to two decimal places if n is between 100 and 999, and so on.

Example 5.2

The number of spikelets in five flowerheads of annual meadow grass are recorded below.

28   16   24   31   27

Calculate the mean number of spikelets per flowerhead.

$$\bar{x} = \frac{\Sigma x}{n} = \frac{126}{5} = 25.2$$

In this example, observations relate to a discrete (non-continuous) variable – a whole number of spikelets; the sampling unit is a flowerhead. Notice, however, that the *mean is a fraction*; it can assume any degree of precision. Therefore, a sample mean is a *continuous variable*, even if the observations in the sample are not. The consequences of this important point are considered in Section 11.2.

In Examples 5.1 and 5.2 the mean is derived from **ungrouped observations**. In cases where a larger number of observations are **grouped** into a frequency table, the calculation of the mean is different. The formula for deriving the mean of grouped data is:

$$\bar{x} = \frac{\Sigma fx}{n}$$

where x is the frequency class, f is the frequency of occurrence in a class and Σ is the 'sum of'.

Example 5.3

Calculate the mean of the grouped shoot length measurements given in Section 3.10.

Using the formula given above, the mean is:

$$\begin{aligned} \bar{x} &= \frac{\begin{aligned}&(1 \times 68) + (2 \times 69) + (4 \times 70) + (7 \times 71) + (11 \times 72) + (15 \times 73) \\ &+ (20 \times 74) + (15 \times 75) + (11 \times 76) + (7 \times 77) + (4 \times 78) + (2 \times 79) + (1 \times 80)\end{aligned}}{1 + 2 + 4 + 7 + 11 + 15 + 20 + 15 + 11 + 7 + 4 + 2 + 1} \\ &= \frac{7400}{100} = 74.00 \, \text{mm} \end{aligned}$$

We said in Section 5.1 that an average is a number which represents all the observations in a sample. Consider the next example.

Example 5.4

The numbers of wood ants captured in seven pitfall traps set overnight in a deciduous woodland are:

25   4   12   9   15   8   202

Calculate the mean number of ants per trap.

$$\bar{x} = \frac{\Sigma x}{n} = \frac{275}{7} = 39.3 \, \text{ants}$$

The calculated value of 39.3 in Example 5.4 scarcely represents all the observations in the sample. It is larger than six of the seven observations and is more than five times smaller than the other. Because the mean takes into account the value of every observation in a sample, it can be greatly distorted by a single exceptional value. Perhaps the seventh pitfall trap was set near to a nest or foraging pathway. When a few exceptional values distort the mean in this way, a *resistant measure of the average*, namely the **median**, may be more appropriate.

5.3   The median – a resistant statistic

The median is the middle observation in a set of observations which have been ranked in magnitude.

Example 5.5

Look again at the distribution of wood ant counts given in Example 5.4. The observations are ranked in increasing order:

4   8   9   **12**   15   25   202
            median

The observation 12 has three observations to the left which are smaller and three to the right which are larger. It is the sample **median**. Notice how it is *resistant* to the extreme observation; the median is 12 if the seventh observation is 20, 202 or 2002. The median is more *representative* of this set as a whole than the mean.

If there is an even number of observations in a sample, there is no middle value. By convention, the middle is taken to be the mean of the values of the middle pair.

Example 5.6

Find the median of the following observations:

9.2   11.5   13.2   19.7   29.4   50.1

The median lies between the 3rd and the 4th observation. It is estimated by:

$$\text{Median} = \frac{13.2 + 19.7}{2} = 16.45$$

The same answer is obtained by halving the *difference* between the two middle observations and adding the result to the lower value or subtracting it from the larger:

$$\text{Median} = 13.2 + \frac{19.7 - 13.2}{2} = 16.45$$

An advantage of the median is that it may be determined even if the values of all observations are not known. This is because the median is an **ordinal statistic**; we need only to know the **ranks** of most of the observations, as we show in the next example.

Example 5.7

In a study of migratory behaviour in birds, 15 homing pigeons are transported 1200 km from their loft in England to a place in France where they are released together. As each bird arrives back at the loft it operates a foot treadle which records the time of arrival automatically on a chart recorder. The following day the observer finds that all pigeons are back in the loft but, unfortunately, the treadle jammed after the arrival of the tenth bird. From the recorder, the return times of the 10 birds are:

16 h 45 min;   17 h 30 min;   18 h 05 min;   18 h 15 min;   19 h 20 min;
19 h 25 min;   21 h 10 min;   21 h 55 min;   22 h 10 min;   23 h 25 min.

What is the average time taken to return?

It is not possible to calculate the mean return time without values for the missing observations. Because we know that the five missing values are all larger than 23 h 25 min, we are able to find the median. The middle observation is the 8th from either end of the distribution, namely 21 h 55 min. There are seven values lower than this, and seven which are larger. It does not matter that we do not know the return times of the slowest five pigeons.

Suppose only 14 pigeons were released, and the final four failed to operate the treadle. Now the mid-point is between the 7th and 8th observations, that is, half way between 21 h 10 min and 21 h 55 min. The difference between them is 45 min. Add half of this, 22.5 min, to the lower time:

21 h 10 min + 22.5 = 21 h 32.5 min. This is the median time.

When there are several observations of the same value near to the median, its calculation is a little more complicated, as the next example shows.

Example 5.8

A biologist investigates the dispersal of the mosquito *Aedes detritus* in a saltmarsh. About 1000 larvae are treated with blue stain (Geimsa) in a bucket and returned to the pool from which they were taken. The stain combines with the insects' protein, and remains through pupation into the adult mosquito. Students on a field course search for 'blue' mosquitoes in 100-m zones downwind of the pool. Fig. 5.1 indicates in a dot diagram how many mosquitoes are found in each zone. What is the median distance moved?

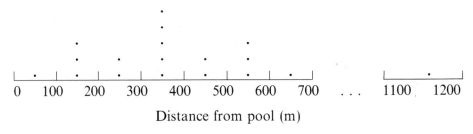

Distance from pool (m)

Fig. 5.1   Positions where mosquitoes marked in the pool are located

1. Calculate the median position: this is $n/2$. There are 18 mosquitoes: the median is therefore the 9th observation.
2. Starting at the left, pass in turn $1 + 3 + 2$, that is, a cumulative frequency of 6, before arriving at the class interval (300–400 m) which contains the 9th – median – observation. If the median is the 9th, and we have already passed 6, the median is the 3rd observation into the interval.
3. There are five observations in that interval. Therefore the median is 3/5ths of the way into the interval.
4. The breadth of the interval is 100 m; 3/5ths of this is 60 m.
5. Add 60 m to the value of the lower limit of the interval: $300 + 60 = 360$ m. This is the median distance.
6. Confirm the result by working down the distribution from the right. There

is a cumulative frequency of 7 before the interval containing the 9th (median) observation is reached. The median is therefore the second observation in the interval. There are five observations in the intervals. Therefore $2/5$ of $100\,\text{m} = 40\,\text{m}$. Because we are working from the right, 40 m is subtracted from the value of the upper limit of the interval, that is $400 - 40 = 360\,\text{m}$.

We can compare this *median* value with that of the *mean* by using the method shown in Example 5.3. In this instance we do not know the actual distance moved by each mosquito, only in which zone it was found. We therefore have to approximate by using the mid-point or class mark of each zone (50, 150, 250 m . . .). The mean is:

$$\frac{(1 \times 50) + (3 \times 150) + (2 \times 250) + (5 \times 350) + (2 \times 450) + (3 \times 550) + (1 \times 650) + (1 \times 1150)}{18}$$

$\bar{x} = 394.4\,\text{m}$

Notice that the mean is larger than the median because, again, a single very large observation distorts the mean in the direction of that large value.

In the example, we place observations within defined class intervals of 100 m zones. The method of calculating the median is similar when observations are measured.

Example 5.9

Calculate the median of the following nine measurements (mm):

2   3   3   4   5   5   5   6   10

1. Draw a dot diagram of the data.

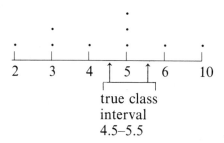

2. The median is the $n/2 = 9/2 = 4.5$, that is, the 4.5th observation from either end of the distribution.
3. Starting from the left, the cumulative frequency in the first three classes is $1 + 2 + 1 = 4$. The median is therefore the 0.5th observation in the next class.

There are three observations in the next class; therefore the median is $0.5/3 = 0.167$th of the way into the class.

4. We recognize that the three observations of 5 mm are not *precise*: they are accurate to ± 0.5 mm. The observations fall within a class interval with implied lower and upper limits of 4.5 and 5.5 mm.
5. The breadth of the class interval is $5.5 - 4.5 = 1$ mm. The median is 0.167th of the way into the interval, that is, 0.167 mm. Add this to the value of the lower interval limit: $4.5 + 0.167 = 4.667$ mm. This, then, is the median.
6. Confirm the result by working down the distribution from the right. The median observation is the $n/2 = 4.5$th observation. The cumulative frequency down to the median class interval is $1 + 1 = 2$. The median observation is therefore the 2.5th into the interval. There are three observations in the interval; the median, then, is $2.5/3 = 0.833$th of the way into the interval. Because the interval breadth is 1 mm, 1×0.833 is subtracted from the value of the upper interval limit, $5.5 - 0.833 = 4.667$ mm. The median checks at 4.667 mm.

5.4　The mode

The mode is another measure of the average. In its most common usage this measure is called the **crude mode** or **modal class**. It is the class in a frequency distribution which contains more observations than any other. The modal shoot length class in Section 3.10 is 74 mm. The mode of the sample of measurements in Example 5.9 is 5 mm. The mode is the only measure of the average that can be applied to observations on *ordinal* scales.

Example 5.10

What is the average *DAFOR* score of the occurrence of *Nardus stricta* (mat grass) in 100 quadrats surveyed on Exmoor?

Dominant:　　15
Abundant:　　28
Frequent:　　35
Occasional:　　12
Rare:　　10

The modal score is *Frequent* (35).

A frequency distribution with more than one peak, or mode, is called a **multimodal distribution**. When there are two peaks it is a **bimodal** distribution. Figure 5.2 shows a frequency distribution of observations of the mass of hedgehogs. There is one mode at the interval represented by 780 g and another at 900 g. In this instance the two modes correspond to the modal masses of males and females. As we explain in Chapter 9, some statistical techniques depend on an assumption that data are approximately distributed in a special symmetrical way in which the mode is on the axis of symmetry. Because multi-

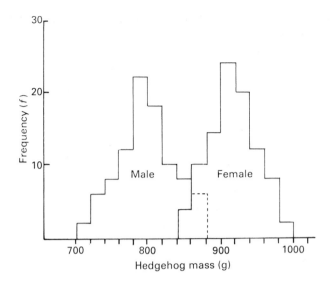

Fig. 5.2   Biomodal distribution of hedgehog masses.

modal distributions are not like this, it may be necessary to conduct separate analyses on discrete population categories (males, females, juveniles, etc.) for which the data are more or less symmetrical. A dot-diagram usually shows if the data are multi-modal (see Example 9.5(c)).

5.5   Relationship between the mean, median and mode

In a perfectly symmetrical distribution the mean, median and mode have the same value. In a skewed distribution, the mean shifts towards the direction of the skew. In biological distributions a skew is nearly always to the right (positive skew) and so the mean is larger than the median and the mode, as Fig. 5.3 illustrates.

The mean is the only one of the three measures of average which uses all the

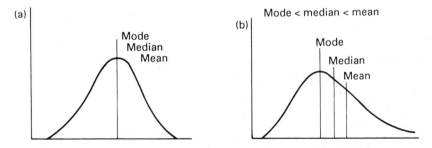

Fig. 5.3   (A) Symmetrical distribution; (B) skewed distribution.

information in a set of data. That is to say, it reflects the value of each observation in the distribution. Its precisely defined mathematical value allows other statistical techniques to be based upon it. Moreover, it has the advantage of being capable of combination with the means of other samples to obtain an overall mean. This may be useful if individuals (sampling units) are rare or are hard to come by.

Example 5.11

One observer reports the mean number of fleas on four dormice to be 3.75; another reports a mean of 6.4 from five dormice. What is the overall mean?

The overall mean is:

$$\frac{(4 \times 3.75) + (5 \times 6.4)}{9} = 5.2$$

The overall mean of 5.2 is likely to be a better estimate of the actual (population) mean than either of the means of the two smaller samples.

As we have seen in Examples 5.4 and 5.8, there are occasions when taking account of the value of every observation in a distribution gives a distorted picture of data. The effect of introducing one or two atypically high or low observations is to pull the mean in the direction of those values. In such cases, the median provides a more realistic description of the centre of the distribution than the mean. The median is especially useful, moreover, in the *preliminary* analysis of data in highlighting overall trends.

The mode is useful as a quick and approximate measure of central tendency and as an indication of the centre of distribution of observations measured on *ordinal* scales.

6 MEASURING VARIABILITY

6.1 Variability

If there were no variability within populations there would be no need for statistics: a single item or sampling unit would tell all that is needed to be known about the population as a whole. It follows therefore that in presenting information about a sample it is not enough simply to give a measure of the *average*; what is also needed is information about **variability** within that sample. A dot-diagram is a qualitative method of assessing the variability in a sample, as Fig. 6.1 shows. For a quantitative analysis of variability within and between samples, we need a mathematically defined measure. The three that we describe in this chapter are the **range, standard deviation** and **variance**.

6.2 The range

The simplest measure of variability in a sample is the range. This indicates the highest and lowest observations in a distribution, and shows how wide the distribution is over which the measurements are spread. For continuous variables, the range is the arithmetic difference between the highest and lowest observations in the sample. In the case of counts or measurements recorded to the nearest whole unit, remember to add 1 to the difference because the range is inclusive of the extreme observations.

The range takes account of only the two most extreme observations. It is therefore limited in its usefulness because it gives no information about how observations are distributed. Are they evenly spread, or are they clustered at one end? The range should be used when observations are too few or too

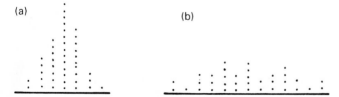

Fig. 6.1 (a) Low variability; (b) high variability.

scattered to warrant the calculation of a more precise measure of variability, or when all that is required is a knowledge of the overall spread of the observations as, for example, when parasitologists express the range of infestation of fleas, lice, etc. on a sample of hosts.

Finding the range is an important step in constructing a frequency distribution (see Section 3.11).

6.3   The standard deviation — measure of variability within a sample.

The standard deviation is the most widely applied measure of variability. It is calculated directly from all the observations of a particular variable. When observations have been obtained from *every* item or sampling unit in a population, the symbol for the standard deviation is σ (sigma). This is a parameter of the population and it is calculated from:

$$\sigma = \sqrt{\frac{\Sigma(x-\mu)^2}{N}}$$

where $x =$ the value of an observation, $\mu =$ population mean, $\Sigma =$ the 'sum of' and $N =$ number of items or sampling units in the population. It is rarely possible to obtain observations from every item or sampling unit in a population. We have therefore to be content with *estimating* σ from the observations in a sample. The estimate of σ is symbolized s. The value of s is not absolute, but varies from sample to sample. However, a set of values of s obtained from several samples clusters around σ. The formula for obtaining s is:

$$s = \sqrt{\frac{\Sigma(x-\bar{x})^2}{n-1}}$$

$n-1 = $ degrees of freedom

where $\bar{x}$ is the sample mean and n is the number of observations in the sample. The term $(x-\bar{x})$ is known as the **deviation from the mean**. In cases where $\bar{x}$ is larger than x the deviation is negative; squaring the number always makes it positive. In the formula for s, note that the denominator is $(n-1)$. The reduction by one has the effect of increasing s. Elliott (1977) describes this increase neatly as a 'tax' to be paid for using the sample mean $\bar{x}$ (a statistic) instead of the population mean μ (a parameter) in the estimation of σ. The expression $(n-1)$ is known as the **degrees of freedom**. We explain the meaning of this expression more fully at the end of the chapter, but in estimating a standard deviation, the degrees of freedom are always one less than the number of observations in a sample. Although s is an estimate of the *population standard deviation*, it is widely referred to as the **sample standard deviation** because it is calculated from sample data.

It is possible to calculate an alternative sample standard deviation by placing n (rather than the degrees of freedom, $n-1$) in the denominator. This, however, is a *descriptive statistic only* and has no meaning in inferential statistics. We recommend that you never use it unless you have a very clear reason for doing so. In practice, you will undoubtedly choose to derive a standard deviation directly from a scientific calculator. This usually has separate keys marked n and $(n-1)$ for population and sample standard deviations, respectively. In Section 6.4 we show how to calculate the standard deviation from first principles. It is important that you study this because it introduces a term called the sum of squares which has application in more advanced statistical methods that we touch upon in later chapters.

6.4 Calculating the standard deviation

The procedure for calculating the standard deviation from first principles is shown in Example 6.1.

Example 6.1

Calculate the standard deviation of the following 10 observations (mm).

81	79	82	83	80	78	80	87	82	82

1. Calculate the mean, $\bar{x}$:

$$\bar{x} = \frac{\Sigma x}{n} = \frac{814}{10} = 81.40 \text{ mm}$$

2. Obtain the deviations from the mean by subtracting the mean from each observation in turn; square each deviation (the squaring eliminates any minus signs):

$(81-81.4)^2 = 0.16$ $(78-81.4)^2 = 11.56$
$(79-81.4)^2 = 5.76$ $(80-81.4)^2 = 1.96$
$(82-81.4)^2 = 0.36$ $(87-81.4)^2 = 31.36$
$(83-81.4)^2 = 2.56$ $(82-81.4)^2 = 0.36$
$(80-81.4)^2 = 1.96$ $(82-81.4)^2 = 0.36$

3. Add up the 10 squared deviations: The sum of the squared deviations is 56.4. This sum, $\Sigma(x-\bar{x})^2$ is called the **sum of squares of the deviations** or, more simply, the **sum of squares**.
4. Divide the sum of squares by one less than the number of observations:

$$\frac{\text{sum of squares}}{(n-1)} = \frac{\Sigma(x-\bar{x})^2}{(n-1)} = \frac{56.4}{9} = 6.27 \; = variance$$

5. The standard deviation is the square root of this value:

$$s = \sqrt{6.27} = 2.50 \text{ mm}$$

We estimate that the sample of 10 observations is drawn from a population whose standard deviation is 2.50 mm.

6.5   Calculating the standard deviation from grouped data

The calculation of the standard deviation of a large sample is less of a chore if the data are grouped. The formula for the standard deviation of grouped data is:

$$s = \sqrt{\frac{\Sigma f(x - \bar{x})^2}{n - 1}}$$

Example 6.2

Calculate the standard deviation of the 100 shoot-length measurements of Section 3.10, of which $\bar{x}$ has already been worked out in Example 5.3 as 74.0 mm.

1. Obtain the value of $(x - \bar{x})^2$ for each frequency class, and multiply it by the number of observations in the class. Thus, in frequency class 68, $(x - \bar{x})^2 = (68 - 74)^2 = 36$; $36 \times 1 = 36$. Sum the columns of f and $f(x - \bar{x})^2$. The results are shown in Table 6.1.

Table 6.1   Calculating a standard deviation of grouped data

Frequency class (x) mm	$(x - \bar{x})^2$	Frequency, f	$f(x - \bar{x})^2$
68	36	1	36
69	25	2	50
70	16	4	64
71	9	7	63
72	4	11	44
73	1	15	15
74	0	20	0
75	1	15	15
76	4	11	44
77	9	7	63
78	16	4	64
79	25	2	50
80	36	1	36
		$n100$	$\Sigma f(x - \bar{x})^2 = 544$

2. $s = \sqrt{\dfrac{\Sigma f(x - \bar{x})^2}{n - 1}} = \sqrt{\dfrac{544}{99}} = \sqrt{5.49} = 2.34 \, \text{mm}$

Scientific calculators have a facility for entering grouped data for computing a standard deviation.

6.6 Variance

An important measure of variability closely related to the standard deviation is variance. Variance is used in certain parametric techniques described later. **Variance** is the square of the standard deviation; conversely, a standard deviation is the square root of the variance. Variance is symbolized σ^2 for a population variance, and s^2 for a variance estimated from a sample. Thus:

$$\sigma = \sqrt{\sigma^2} \quad \text{and} \quad s = \sqrt{s^2} \quad \text{and} \quad s^2 = \frac{\Sigma(x-\bar{x})^2}{n-1}$$

It follows that the variance is the value obtained before taking the square root in the final step of the calculation of the standard deviation. The variance of the sample in Example 6.1 is 6.27 and in Example 6.2 is 5.49.

6.7 An alternative formula for calculating the variance and standard deviation

The methods described in Sections 6.4 and 6.5 illustrate the principle of calculating the variance and standard deviation. An algebraic rearrangement of the formula is, in practice, easier to handle. Moreover, it introduces two statistical expressions which will appear again later.

The alternative formula for obtaining the sum of squares is:

$$\text{Sum of squares} = \Sigma x^2 - \frac{(\Sigma x)^2}{n}$$

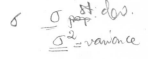

$\sigma \quad$ st. dev.

σ^2 = variance

The two new expressions are Σx^2 and $(\Sigma x)^2$

(i) Σx^2 is called *the sum of the squares of x*. Using again the 10 measurements of Example 6.1 it is derived as follows:

| x: | 81 | 79 | 82 | 83 | 80 | 78 | 80 | 87 | 82 | 82 |

$$\Sigma x^2: \quad 81^2 + 79^2 + 82^2 + 83^2 + 80^2 + 78^2 + 80^2 + 87^2 + 82^2 + 82^2 = 66\,316$$

(ii) $(\Sigma x)^2$ is called the *square of the sum of x* and is calculated as follows:

$$(\Sigma x^2) = (81 + 79 + 82 + 83 + 80 + 78 + 80 + 87 + 82 + 82)^2 = (814)^2 = 662\,596$$

Substituting in the formula for the sum of squares:

$$\text{Sum of squares} = 66\,316 - \frac{662\,596}{10} = 56.4$$

(i) The variance is obtained by dividing the sum of squares by $(n-1)$:

$$s^2 = \frac{56.4}{(10-1)} = 6.27$$

(ii) The standard deviation is the square root of the variance:

$$s = \sqrt{6.27} = 2.50 \, \text{mm}.$$

6.8   Obtaining the standard deviation, variance and the sum of squares from a calculator

Scientific calculators have facilities for entering observations and, by pressing appropriate keys, obtaining $\bar{x}$, standard deviation (remember to use the key marked $n-1$), Σx, and Σx^2.

If you are unsure about obtaining these numbers from your calculator, we recommend that you stop at this point and learn to do so from the instruction booklet accompanying the instrument before proceeding further.

Calculators do not generally have facilities for obtaining the variance, sum of squares and $(\Sigma x)^2$ directly.

(i) To obtain the variance, first obtain the standard deviation, and then square it.
(ii) To obtain $(\Sigma x)^2$, obtain Σx and then square it.
(iii) To obtain the sum of squares, obtain the variance as in (i) and then multiply by $(n-1)$.

These three operations can be undertaken while the calculator is operating in the *standard deviation mode*.

Some makes of calculator do not permit the direct retrieval of a standard deviation for $(n-1)$. This is unfortunate because almost without exception this is what you need. To convert s_n to $s_{(n-1)}$, square the standard deviation, multiply by n, divide by $(n-1)$ and take the square root. Check the result of your attempt against the solution to Example 6.1.

6.9   Degrees of freedom

In obtaining a sample standard deviation to estimate a population standard deviation in Section 6.3, reference is made to the number of *degrees of freedom*. Because the concept of degrees of freedom is involved in many statistical techniques, it now needs a fuller explanation.

Suppose that we are told that a sample of $n=5$ observations has a mean $\bar{x}$ of 50 and we are then asked to 'invent' the values of the observations. We know that $\Sigma x = (\bar{x} \times n) = 250$. If the sum of the observations is 250, we have freedom of choice only for the first four observations because the 5th must be a number (perhaps a negative number) that brings the sum to 250. By way of example, if the first four numbers are arbitrarily chosen as, say, 40, 25, 18, 130, then in order to make $\Sigma x = 250$, we have no further freedom of choice; the fifth number must be 37. Degrees of freedom (df) in our present example is one less than the number of observations. That is, $df = (n-1)$.

Sometimes the formula for estimating a population parameter contains a value which is itself an estimate. Thus, to estimate a standard deviation, a knowledge of the mean is required. Because the value of the mean is itself estimated from a sample, this 'costs' a degree of freedom (this cost is Elliott's 'tax' referred to in Section 6.3).

The degrees of freedom are not always $(n-1)$; they depend on the particular estimation in hand. We explain the rule for deciding the degrees of freedom for each technique as it arises. Degrees of freedom are symbolized by v (nu).

6.10 The coefficient of variation CV

The standard deviation s is a measure of the degree of variability in a sample which estimates the corresponding parameter of a population. However, it is of limited value for *comparing* the variability of samples whose means are appreciably different. The standard deviation of the mass of a population of blue whales is hundreds of kilograms whilst that of a population of harvest mice is a few grams. When comparing variability in samples from populations with different means, the **coefficient of variation (CV)** is used. This is the ratio of the standard deviation to the mean, usually expressed as a percentage by multiplying by 100.

Example 6.3

Aspects of growth in bird chicks are investigated. The mean and standard deviation of samples of (a) eggs; (b) 4-day-old chicks; (c) 10-day-old chicks are recorded. Does the relative variability change with age?

(a) *Eggs:* $\bar{x}=3\,g$; $s=0.54\,g$. $CV=0.54/3 \times 100=18\%$.
(b) *4-day-old chicks:* $\bar{x}=4.5\,g$; $s=1.3\,g$. $CV=1.3/4.5 \times 100=28.9\%$.
(c) *10-day-old chicks:* $\bar{x}=10.4\,g$; $s=4.1\,g$. $CV=4.1/10.4 \times 100=39.4\%$.

It is clear from an inspection of the three values of CV that relative variability increases with age.

7 PROBABILITY

7.1 The meaning of probability

The mathematical theory of probability arose from the study of games of chance. That is why so many text-book examples of probability involve dice-throwing or coin-tossing. Probability may be regarded as quantifying the chance with which a stated outcome of an event will take place. By convention, probability values lie on a scale between 0 (impossibility) and 1 (certainty) but they are sometimes expressed as percentages. Some examples will make this clear.

Example 7.1

Long-term studies of a pond community have shown that it is inhabited by only two species of meniscus midge (Diptera: Dixidae), namely *Dixella autumnalis* and *Dixella attica*. The pupae, which are indistinguishable and are dispersed randomly and independently in the habitat, are collected from emergent vegetation just above the water line, and emerge as adults within a few days in captivity. On the basis of very large samples it is known that on average 80 out of every 100 pupae collected will emerge as *D. autumnalis* and the remainder as *D. attica*. It is also known that, on average, insects emerge in the ratio 1 : 1 males to females, i.e. 1 out of every 2 is a male, and the other a female. We will assume, for illustrative purposes, that there is negligible mortality during pupation, and that an 'event' is the hatching of a pupa which has been selected at random.

(a) What is the probability that the outcome of an event is *D. autumnalis*?

$$P = \frac{\text{Number of nominated outcomes}}{\text{Total number of possible outcomes}}$$

$$P = \frac{\text{Number of pupae emerging as } D.\ autumnalis}{\text{Total number of pupae emerging}} = \frac{80}{100} = 0.8\ (80\%)$$

Notice that the probability value is equal to the *proportional frequency* of the species in a large sample.

(b) What is the probability that the outcome of an event is *D. attica*? Similarly:

$$P = \frac{\text{Number of pupae emerging as } D. \; attica}{\text{Total number of pupae emerging}} = \frac{20}{100} = 0.2 \; (20\%)$$

(c) What is the probability that the outcome of an event will be a male?

$$P = \frac{\text{Number of pupae emerging as males}}{\text{Total number of pupae emerging}} = \frac{1}{2} = 0.5 \; (50\%)$$

7.2 Compound probabilities

Probability values may be added and multiplied (they may also be subtracted and divided but the practical application of these is rare). Because probability values are fractions, adding probabilities *increases* the likelihood of a stated outcome whilst multiplying probabilities *decreases* it. The decision whether to add or to multiply probabilities may often therefore be made intuitively.

Example 7.2

Using the same material as in Example 7.1, estimate the probability that the outcome of an event will be either *D. autumnalis* or *D. attica*.

Clearly, the probability of this outcome is greater than that in which a single species is nominated:

Probability of outcome *D. autumnalis* = 0.8
Probability of outcome *D. attica* = 0.2
Probability of outcome *D. autumnalis* or *D. attica* = 0.8 + 0.2 = 1.0

Example 7.3

What is the probability that the outcome of an event will be female *D. attica*?

In this case, since two conditions (species *and* sex) are stipulated, the probability of the outcome *female D. attica* is lower than for the outcome *D. attica* alone:

Probability of outcome *D. attica* = 0.2
Probability of outcome female = 0.5
Probability of outcome female *D. attica* = 0.2 × 0.5 = 0.1

That is, on average, one pupa in ten will emerge as a female *D. attica*.

Example 7.4

If the outcome of an event is described in terms of species and sex, show that the sum of the probabilities for all possible outcomes is equal to 1.0.

	Probability of outcome		
Male D. autumnalis	*Female* D. autumnalis	*Male* D. attica	*Female* D. attica
$0.8 \times 0.5 = 0.4$	$0.8 \times 0.5 = 0.4$	$0.2 \times 0.5 = 0.1$	$0.2 \times 0.5 = 0.1$

Total probability $= 0.4 + 0.4 + 0.1 + 0.1$
$\qquad\qquad\quad = 1.0$

A breakdown which shows the individual probabilities of all possible outcomes and which adds up to 1.0 is called a **probability distribution**.

7.3   Probability distribution

Section 7.2 concludes with an example of a simple probability distribution. Probability distributions may be generated empirically, that is, by sampling and observation. Thus, recalling Fig. 4.7 which is a frequency distribution of a sample of 100 observations of shoot lengths, we can convert a frequency distribution into a probability distribution by re-scaling the vertical axis and dividing by the number of observations (sample size). This has been done in Fig. 7.1. It may be seen that, for example, the probability that a shoot selected at random will have a length (when rounded to a whole number) of

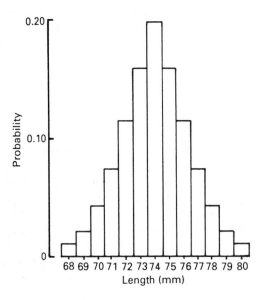

Fig. 7.1   Probability distribution of leaf length (± 0.5 mm) estimated from a sample of 100 leaves described in Section 3.10.

74 mm = 20 ÷ 100 = 0.2. Moreover, if the individual probabilities for each frequency class are summed, the result is 1.0.

Why are probability distributions useful? First, as indicated above, they allow us to estimate the probability that an event (e.g. the random selection of a shoot) will have a stated outcome (e.g. length 74 mm when rounded). Second, a probability distribution may be used to generate a distribution of 'expected frequencies'. Just as a probability is estimated by dividing a frequency of a particular observation by the total number of all observations, an expected frequency may be produced by *multiplying* a probability by the number of observations:

Expected frequency = (estimated probability) × (number of observations)

Example 7.5

Using the probability distribution estimated in Example 7.4, what is the *expected* frequency of each outcome in terms of species and sex in a sample size of 200 collected from a different pond if the same conditions are presumed to apply?

Expected frequency of male *D. autumnalis*   = 0.4 × 200 = 80
Expected frequency of female *D. autumnalis* = 0.4 × 200 = 80
Expected frequency of male *D. attica*   = 0.1 × 200 = 20
Expected frequency of female *D. attica*   = 0.1 × 200 = 20

Total expected frequency   = 200

In practice, the obvious thing to do next is to compare the expected frequencies of each outcome with actual or observed frequencies. If the observed frequencies disagree with those expected, there could be a basis for exploring factors which underlie the discrepancy.

7.4   Models of probability distribution

So far we have illustrated examples of probability and probability distribution which are estimated from sample data. There is another way of generating a probability distribution, however. That is according to a mathematical prescription, or theorem, often referred to as a 'mathematical model'. Mathematical modelling has considerable application in experimental investigation. If it can be shown that certain types of data (say, a frequency distribution) agree well with a distribution predicted by a mathematical model, we may begin to make generalizations about such data. On the other hand, if experimental data do *not* agree with those predicted, we may have doubts about our initial hunch and further experimental field work might be called for.

There are a number of mathematical models available for generating probability distributions, but only a few are of practical importance to field biologists. In the remainder of this chapter we examine three distributions which are associated with *discrete* variables (that is, *counts* of things): **binomial**, **Poisson** and **negative binomial**. We then illustrate some of their practical applications. In Chapter 9 we introduce probability distributions which are appropriate for *continuous* variables (that is, *measurements*).

7.5   The binomial probability distribution

The binomial distribution is appropriate when:

(a) observations consist of counts of things
(b) an observation is classified into one of two possible categories. Examples are:

   male *or* female
   species A *or* species B
   species A *or* not species A
   success *or* failure (e.g. seed germinating; egg hatching)
   adult *or* not adult

We will show later that the binomial distribution is also appropriate when:

(c) the variance of a sample of count data is less than the mean
(d) objects being counted are dispersed in a *regular* fashion.

We explore the properties of the binomial distribution by considering a situation in which there are only two possible outcomes from a single event or trial and the probability of each outcome is equal. If we let p be the probability of one outcome and q be the probability of the other, then $p+q=1$; $q=(1-p)$ and $p=0.5=q$. Such is the case for the insect illustrated in Example 7.1 which from long term studies is known to emerge in equal numbers of each sex. We let the number of events (number of pupae taken) $=k$, the probability of a male $=p$ and the probability of a female $=q$. The possible outcomes of a single event $(k=1)$ are shown in Table 7.1.

Table 7.1   Binomial distribution, single event

No. of events (k)	Possible outcomes $=2$		No. of combinations	Total probability
1	M	F	2	
	1/2	1/2		$1/2+1/2=1$
	p	q		$p+q \quad =1$

If 2 pupae are taken $(k=2)$ the number of possible outcomes increases to four: 2 males (MM); male followed by female (MF); female followed by male (FM); and finally, 2 females (FF). Because there are four possible outcomes, the probability of *any one* is 1 in 4 or 0.25. However, since MF and FM are the same combination, differing only in which comes first, the number of possible combinations is three. This is summarized in Table 7.2.

Similarly, if 3 pupae are taken ($k=3$), the number of possible outcomes increase to 8, and the possible combinations to four (Table 7.3). This can be expanded, with slight abbreviation, for 4 pupae ($k=4$) in Table 7.4. We could continue to build up tables for greater numbers of events, that is, larger values of k but it would be cumbersome to do so. Instead, let us pause here to make some generalizations about what has emerged already.

1. Note that we have an example of compound probabilities in which probabilities have been both added and multiplied as described in Example 7.4. Thus, referring to Table 7.3 in which $k=3$, the probability that the outcome of any *prescribed* sequence of three events (e.g. MMM, MFM, FFM, etc.) is a *product* of probabilities, namely $0.5 \times 0.5 \times 0.5 = 0.125$. However, if the requirement of a prescribed sequence is removed, and we seek merely the probability of a particular *combination*, say 2 males and 1 female, the probability is the *sum* of the probabilities of the three individual outcomes, namely $0.125 + 0.125 + 0.125 = 0.375$.

2. Whatever the value of k, the sum of the probabilities for each combination can be calculated either by adding up the probabilities of all possible outcomes (e.g. for $k=3$ this is $1/8 + 3/8 + 3/8 + 1/8$) or by application of a mathematical formula in terms of p and q. In either case the total probability is 1. This means that each table ($k=1$ to 4 in our examples) is itself a complete probability distribution according to our definition in Example 7.4. Furthermore, it reveals that there is no such thing as a *single binomial probability distribution*. Rather, by keeping p constant (0.5 in this case) we can build up a 'family' of distributions whose structure is determined by the factor k. Such a factor, whose numerical value determines the distribution of probability, is called a *parameter*. We observe that when $p=0.5=q$, the distribution is symmetrical whatever the value of k. In our example we have controlled the value of k. Sometimes, however, it is necessary to estimate k from a sample. We explain how to do this in Chapter 8.

3. For each value of k, we can describe the distribution of probability for each combination according to a formula in terms of p and q. Thus:

For $k=1$: $p+q=1$
For $k=2$: $p^2 + 2pq + q^2 = 1$
For $k=3$: $p^3 + 3p^2q + 3pq^2 + q^3 = 1$
For $k=4$: $p^4 + 4p^3q + 6p^2q^2 + 4pq^3 + q^4 = 1$

Readers with mathematical knowledge will recognize the development as belonging to successive terms in the expansion of the binomial series $(q+p)^k$.

For practical purposes it would be helpful to have a generalized formula which, for given values of k and p, would allow us to compute the probability of *any* combination. Such a formula exists:

$$P_{(x)} = \frac{k!}{x!(k-x)!} \times p^x \times q^{(k-x)}$$

Table 7.2 Binomial distribution, two events

No. of events (k)	Possible outcomes=4			No. of combinations	Total probability
2	MM (2M=1/4) $p \times p$ $= p^2$ $= 0.25$	MF FM (1M,1F = 1/4+1/4 = 2/4) $pq + qp$ $= 2pq$ $= 0.5$	FF (2F,1/4) $q \times q$ $= q^2$ $= 0.25$	3	$1/4 + 2/4 + 1/4 = 1$ $p^2 + 2pq + q^2$ $= 1$

Table 7.3 Binomial distribution, three events

No. of events (k)	Possible outcomes=8			No. of combinations	Total probability
3	MMM (3M,1/8) $p \times p \times p$ $= p^3$ $= 0.125$	MMF MFM FMM (2M,1F = 3/8) $p^2q + pqp + qp^2$ $= 3p^2q$ $= 0.375$	FFM FMF MFF (2F,1M = 3/8) $q^2p + qpq + pq^2$ $= 3pq^2$ $= 0.375$ FFF (3F,1/8) $q \times q \times q$ $= q^3$ $= 0.125$	4	$1/8 + 3/8 + 3/8 + 1/8$ $= 1$ $p^3 + 3p^2q + 3pq^2 + q^3$ $= 1$

Table 7.4 Binomial distribution, four events

No. of events (k)	Possible outcomes = 16					No. of combinations	Total probability
4	MMMM (4M, 1/16)	MMMF MMFM MFMM FMMM (3M,1F, 4/16)	MMFF MFMF MFFM FMMF FMFM FFMM (2M,2F, 6/16)	FFFM FFMF FMFF MFFF (3F,1M, 4/16)	FFFF (4F, 1/16)	5	$1/16 + 4/16 + 6/16 + 4/16 + 1/16 = 1$
	$p \times p \times p \times p$	$p^3q + p^2qp + pqp^2 + qp^3$	$p^2q^2 + pqpq + pq^2p + qp^2q + qpqp + q^2p^2$	$q^3p + q^2pq + qpq^2 + pq^3$	$q \times q \times q \times q$		$p^4 + 4p^3q + 6p^2q^2 + 4pq^3 + q^4 = 1$
	$= p^4$ $= 0.0625$	$= 4p^3q$ $= 0.25$	$= 6p^2q^2$ $= 0.375$	$= 4pq^3$ $= 0.25$	$= q^4$ $= 0.0625$		

The symbol ! in a formula means 'factorial'; 5! is read '5-factorial'. It means, simply, $5 \times 4 \times 3 \times 2 \times 1 = 120$. Sometimes we need zero-factorial (0!) – this is equal to 1. Although the formula looks a little fearsome it is easy to apply with the help of a calculator. Let us go through the terms:

P = probability of a particular combination
k = number of events or trials
x = a stated number of a particular outcome (e.g. male)
p = probability of a particular outcome (e.g. male)
q = probability of the other outcome (e.g. female)

Whilst field biologists do not need to be burdened with the mathematical proof of the formula, they are certainly entitled to be convinced that it works! We will apply it to solve a problem for which we already know the answer.

Example 7.6

If the probability of a male insect emerging from a pupa selected at random is 0.5, what is the probability that 4 pupae will result in 1 male and 3 females?

(Referring to Table 7.4 where $k = 4$ we see that the probability for 3F,1M is $4/16 = 0.25$.) Checking our values:

$k = 4 \qquad x = 1$ male $\qquad p = 0.5 \qquad q = (1 - p) = 0.5$

Substituting in the general binomial formula

$$P = \frac{4!}{1!(4-1)!} \times 0.5^1 \times 0.5^{(4-1)}$$

$$= \frac{4 \times 3 \times 2 \times 1}{1 \times (3 \times 2 \times 1)} \times 0.5 \times 0.125$$

$$= \frac{24}{6} \times 0.0625$$

$$= 0.25$$

The formula works!
   Now that we have confidence in the formula, let us apply it to solve a more ambitious problem.

Example 7.7

Compute the complete probability distribution of all combinations for $k = 8$ when $p = 0.5$.

Since $p = 0.5$ (and hence $q = 0.5$) let us retain our image of male and female insects to keep the example concrete. If $k = 8$, the possible outcomes are:

male:	8	7	6	5	4	3	2	1	0
female:	0	1	2	3	4	5	6	7	8

There are nine possible combinations in which the sum of males and females is eight. (*Hint*: since we have observed that when $p=0.5=q$ the distribution is symmetrical, we need only compute the probabilities for the first five combinations – the rest follow by symmetry.)

In the outline that follows, remember that only the value of x, the number of males, changes from step to step:

1. For 8 males (also for 8 females):

$$P=\frac{8!}{8!(8-8)!}\times 0.5^8 \times 0.5^{(8-8)}=1\times 0.5^8 \times 1 \qquad\qquad =0.00391$$

2. For 7 males (also for 7 females):

$$P=\frac{8!}{7!(8-7)!}\times 0.5^7 \times 0.5^1 =\frac{40\,320}{5040\times 1}\times 0.00781 \times 0.5 =0.03$$

3. For 6 males (also for 6 females):

$$P=\frac{8!}{6!(8-6)!}\times 0.5^6 \times 0.5^2 =\frac{40\,320}{720\times 2}\times 0.0156 \times 0.25 =0.1092$$

4. For 5 males (also for 5 females):

$$P=\frac{8!}{5!(8-5)!}\times 0.5^5 \times 0.5^3 =\frac{40\,320}{120\times 6}\times 0.031 \times 0.125 =0.217$$

5. For 4 males (also for 4 females):

$$P=\frac{8!}{4!(8-4)!}\times 0.5^4 \times 0.5^4 =\frac{40\,320}{24\times 24}\times 0.0625 \times 0.0625 = 0.2734$$

We can now write out the complete distribution:

No. of males:	8	7	6	5	4	3	2	1	0
No. of females:	0	1	2	3	4	5	6	7	8
Probability:	0.00391	0.03	0.1092	0.217	0.2734	0.217	0.1092	0.03	0.00391

If our computations are correct, the sum of the probabilities should equal 1, subject to small rounding errors during calculation. Please check that they do!

The above probability distribution for $k=8$ and $p=0.5$ is displayed graphically in Fig. 7.2.

Figure 7.2 is revealing in several ways. It shows clearly the symmetry of the distribution when $p=0.5=q$ and indicates that the single most likely outcome from eight events is a combination of 4 males and 4 females ($P=0.2734$), However, by far the most likely outcome from 8 pupae is 5 of one unprescribed sex and 3 of the other. That is, the sum of the probability of 5 males and 3 females *plus* that for 3 males and 5 females: $0.217+0.217=0.43$. There is nearly an 'evens' chance that 8 pupae will generate 5 of one sex and 3 of the other.

In addition to helping us work out the probabilities of various outcomes, the histogram is informative about the nature of binomial distributions in general.

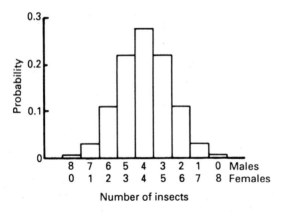

Fig. 7.2　Binomial probability distribution for which $k=8$ and $p=0.5=q$. The height of each bar is the probability of each combination of male and female insects emerging from 8 pupae selected at random.

Note that the x-axis (horizontal) is scaled in integral whole (discrete) numbers. As k becomes large, the increments between the 'steps' in the histogram become progressively smaller, though they will never, in theory, converge to a smooth curve. The y-axis (vertical) is a continuous scale between 0 and a theoretical maximum of 1.

To this point in our outline we have considered the binomial distribution in examples where $p=0.5=q$. This does not have to be the case. Recall Examples 7.2 and 7.3. There, we estimated that the probability of the outcome '*D. autumnalis*' is $p=0.8$. Therefore the probability of the outcome '*D. attica*' is $q=(1-p)=0.2$ (i.e. $p+q=0.8+0.2=1$). Let us use this situation to illustrate a case where $p \neq q$.

Example 7.8

Compute the probability distribution of all outcomes of a binomial distribution where $k=8$ and $p=0.8$.

The procedure is exactly the same as in Example 7.7. To avoid tedious repetition, we illustrate the details of the computation for a single combination, namely where x (the prescribed number of *D. autumnalis*) is 6 (and therefore the number of *D. attica* is 2).

Substituting in the general binomial formula for $x=6$, $p=0.8$ and $q=0.2$ we get:

$$P = \frac{8!}{6!(8-6)!} \times 0.8^6 \times 0.2^2 = 0.294$$

The probabilities of all combinations are set out in the table below. We suggest that you verify the calculations and confirm that the sum of the probabilities is 1.00 (subject to small rounding errors).

No. of *D. autumnalis*:	8	7	6	5	4
No. of *D. attica*:	0	1	2	3	4
Probability:	0.168	0.336	0.294	0.147	0.0459

No. of *D. autumnalis*:	3	2	1	0
No. of *D. attica*:	5	6	7	8
Probability:	0.00918	0.00147	0.000082	0.00000256

The distribution is shown graphically in Fig. 7.3. Notice the skew, a feature of distributions in which $p \neq q$. We can readily see that the most likely outcome from 8 pupae is 7 *D. autumnalis* and 1 *D. attica*. Nevertheless, there is a rather greater than 'evens' chance that 8 pupae will produce either 7 or 6 *D. autumnalis* $(0.336 + 0.294 = 0.63)$.

We could if we wished retain $k = 8$ and change the value of p at will to

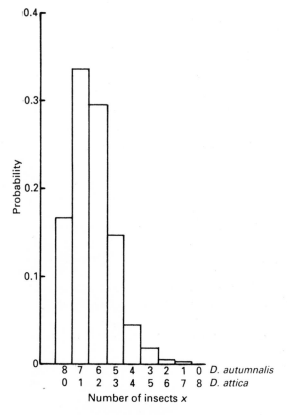

Fig. 7.3 Binomial probability distribution for which $k = 8$ and $p = 0.8$. The height of each bar is the probability of each combination of *D. autumnalis* and *D. attica* emerging from 8 pupae selected at random.

generate a new 'family' of distributions and their histograms. We would find that the distribution tends to symmetry when p and q are nearly equal (e.g. 4.5 and 5.5) and becomes progressively more skewed as their values become more extreme (e.g. 0.95 and 0.05). The value p is therefore also a parameter of the distribution. The value of q is not given the status of a parameter because its value is dependent on p.

Provided that k is a whole number and p is between 0 and 1, the parameters k and p can assume any value, leading to an infinite gradation of probability distributions. But before any *particular* distribution can be described we have to know what are the values of k and p. We say therefore that *the binomial distribution is determined by the parameters k and p.*

7.6   The Poisson probability distribution

The Poisson probability distribution is appropriate when:

(a) observations consist of counts of things;
(b) counts are obtained from defined sampling units, for example, quadrats, intervals of time, etc. and can be organized into a frequency distribution whose mean is approximately equal to the variance;
(c) objects being counted are 'rare'; that is to say, there are far fewer objects in the sampling unit than it is capable of containing;
(d) individuals are dispersed randomly in space or time; they are independent of each other, that is, individuals are not attracted to nor repelled by each other.

There are several restrictions associated with the distribution. Indeed, genuine Poisson distributions are hard to find in nature. Why, then, should we bother? The answer is that it is the *only* model available that is suitable for describing objects which are approximately randomly dispersed. Classical examples include *events* rather than objects, such as radioactive disintegrations per unit time, lightning strikes per unit time and goals in football matches (Moroney, 1956). Lone migrating seabirds passing a headland in unit time intervals and blood cells in unit area of a haemocytometer are also likely to constitute a Poisson distribution.

Recall that the binomial probability distribution is determined by two parameters, k and p. The Poisson distribution is determined by a single parameter called lambda (λ). It can be shown mathematically that λ is equal to the population mean, μ, and also to the variance, σ^2:

$$\mu = \lambda = \sigma^2$$

Since these parameters are *population* characteristics, we are usually obliged to estimate them from sample data, normally from the sample mean $\bar{x}$.

The general formula for a Poisson distribution allows us to estimate the probability, P, that a sampling unit will contain a specified number (x) of individuals for a given value of $\bar{x}$:

$$P_{(x)} = e^{-\bar{x}} \times \frac{\bar{x}^x}{x!}$$

where e is a mathematical constant equal to 2.7183, $\bar{x}$ is the sample mean (an estimate of λ) and x is a specified number of individuals in a sampling unit.

Example 7.9

Estimate the Poisson probability distribution for $x=0$ to 10 when $\bar{x}$ (the estimate of λ) is 4.0.

In other words, we wish to calculate the probability of finding a given number, x, of individuals in a sampling unit where x can assume a whole number between 0 and 10. The values we need are $\bar{x}=4.0$, e $=2.7183$ and $x=0$ to 10 in turn.

For $x=0$ (remember that $0!=1$):

$$P_{(x=0)} = 2.7183^{-4} \times \frac{4^0}{0!}$$

$$= 0.0183$$

(*Hint*: the term $e^{-\bar{x}}$ ($=0.0183$ in this example) occurs in all subsequent steps and its value can usefully be entered in the memory of your calculator.)

For $x=1$:

$$P_{(x-1)} = 2.7183^{-4} \times \frac{4^1}{1!} = 0.0183 \times 4 = 0.073$$

For $x=2$:

$$P_{(x=2)} = 2.7183^{-4} \times \frac{4^2}{2!} = 0.0183 \times 8 = 0.146$$

For $x=3$:

$$P_{(x=3)} = 2.7183^{-4} \times \frac{4^3}{3!} = 0.0183 \times 10.67 = 0.195$$

and so on to $x=10$.

Complete the remaining calculations to verify that:

$P_{(x=4)}=0.195$; $P_{(x=5)}=0.156$; $P_{(x=6)}=0.104$; $P_{(x=7)}=0.059$; $P_{(x=8)}=0.030$; $P_{(x=9)}=0.013$; $P_{(x=10)}=0.005$

The sum of the probabilities for $x=0$ to 10 is 0.994. This is less than 1 and is due to the fact that by stopping at $x=10$ we have not completed the distribution. However, the probabilities for values of x larger than 10 (in this example) become increasingly minute and may be ignored. The probability distribution we have estimated for $\bar{x}=4.0$ is shown in Fig. 7.4. It has a skew to

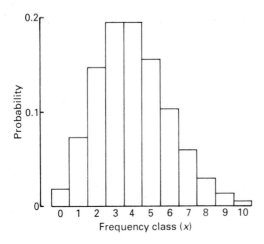

Fig. 7.4 Poisson probability distribution estimated from $\bar{x}=4.0$. The height of each bar is the estimated probability that a sampling unit selected at random will contain x individuals.

the right. The skew becomes more pronounced as the value of λ (and hence its estimate $\bar{x}$) becomes smaller. When λ exceeds 10 the distribution is nearly symmetrical (see also Fig. 8.4).

7.7   The negative binomial probability distribution

The negative binomial distribution is appropriate when:

(a) observations consist of counts of things;
(b) counts are obtained from defined sampling units, for example, quadrats, standardized nets or traps, etc. and can be organized into a frequency distribution whose variance is considerably larger than the mean;
(c) individuals are not dispersed randomly nor are they independent, but may exhibit clumping.

Unlike the Poisson distribution, the negative binomial is a robust model which can describe a wide range of natural situations, including for example: the dispersal of plants in a field; exit holes of the common furniture beetle (woodworm); ectoparasites on hosts (an individual host being a sampling unit); insect eggs on a substrate; spangle galls on oak leaves.

The negative binomial probability distribution is defined in terms of two parameters, μ (the population mean) and an exponent k. Unlike the binomial distribution, in which k is a whole number, the value of k is a continuous variable in the negative binomial.

In practice, the two parameters have to be estimated from a sample. The sample mean, $\bar{x}$, is a satisfactory estimate of μ, whilst k may be estimated from the sample mean and variance according to the relationship:

$$\hat{k} = \frac{\bar{x}^2}{(s^2 - \bar{x})}$$

The cap $\hat{\ }$ over the k indicates it is an estimate of the population parameter.

This is a rather crude estimate of k. There are more accurate methods available but they are considerably more complicated. For practical purposes, the crude estimate is satisfactory.

The individual probabilities of $P_{(x)}$, that is, the probability that a specified number of individuals, x, will occur in a sampling unit selected at random, are derived from an expansion of the expression $(q-p)^{-k}$, where $p = \mu/k$ and $q = (1+p)$. However, the formula for calculating the probability for an individual value of x is so complex that we fear that it would be off-putting to many biologists if we were to print it!

Instead, we will show you an easier method which requires a calculator that has a memory. The price of the simplification is that we cannot pick a value of x at will but must commence at $x = 0$ and work successively through 1, 2, 3, etc. until we wish to stop. This is no disadvantage because, almost always, these low values of x represent important frequency classes in the negative binomial distribution.

Example 7.10

Estimate the negative binomial distribution for $x = 0$ to 5 when $\bar{x}$ of a sample is 2.0 and $s^2 = 5.0$.

We are given the estimate of one parameter (μ), namely 2.0. We estimate the other k, from:

$$\hat{k} = \frac{\bar{x}^2}{(s^2 - \bar{x})} = \frac{2.0^2}{(5.0 - 2.0)} = \frac{4}{3} = 1.33$$

Step 1
The probability for $x = 0$ is given by:

$$P_{(x=0)} = \left(1 + \frac{\bar{x}}{\hat{k}}\right)^{-\hat{k}} = \left(1 + \frac{2.0}{1.33}\right)^{-1.33} = 0.295 \qquad \text{[Store 0.295 in memory]}$$

Step 2
Calculate a factor **F** from:

$$\mathbf{F} = \frac{\bar{x}}{\bar{x} + \hat{k}} = \frac{2.0}{2 + 1.33} = 0.600$$

Enter this number as a constant in your calculator, or write it down, because you need it in all subsequent steps of the calculation.

Step 3

The probability for $x = 1$ is given by:

$$P_{(x=1)} = \frac{\hat{k}}{1} \times F \times P_{(x=0)}$$

$$= \frac{1.33}{1} \times 0.600 \times 0.295 \qquad \text{[Recall 0.295 from memory then clear memory]}$$

$$= 0.235 \qquad \text{[Store 0.235 in memory]}$$

Step 4

The probability for $x = 2$ is given by:

$$P_{(x=2)} = \frac{(\hat{k}+1)}{2} \times F \times P_{(x=1)}$$

$$= \frac{(1.33+1)}{2} \times 0.600 \times 0.235 \qquad \text{[Recall 0.235 from memory then clear memory]}$$

$$= 0.164 \qquad \text{[Store 0.164 in memory]}$$

Step 5

The probability for $x = 3$ is given by:

$$P_{(x=3)} = \frac{(\hat{k}+2)}{3} \times F \times P_{(x=2)}$$

$$= \frac{(1.33+2)}{3} \times 0.600 \times 0.164 \qquad \text{[Recall 0.164 from memory then clear memory]}$$

$$= 0.1092 \qquad \text{[Store 0.1092 in memory]}$$

Step 6

The probability for $x = 4$ is given by:

$$P_{(x=4)} = \frac{(\hat{k}+3)}{4} \times F \times P_{(x=3)}$$

$$= \frac{(1.33+3)}{4} \times 0.600 \times 0.1092 \qquad \text{[Recall 0.1092 from memory then clear memory]}$$

$$= 0.0709 \qquad \text{[Store 0.0709 in memory]}$$

Step 7

The probability for $x = 5$ is given by:

$$P_{(x=5)} = \frac{(\hat{k}+4)}{5} \times F \times P_{(x=4)}$$

$$= \frac{(1.33 \times 4)}{5} \times 0.600 \times 0.0709 \qquad \text{[Recall 0.0709 from memory then clear memory]}$$

$$= 0.0453 \qquad \text{[Store 0.0453 in memory]}$$

By now you will have noticed a pattern building up. The probability for an individual value of x is estimated by multiplying the probability already calculated for the previous value of x (i.e. $x-1$), by a constant term which is calculated in Step 2, and then by another term which, for a given value of x, is:

$$\frac{\hat{k}+(x-1)}{x}$$

We summarize the probability values thus far estimated:

$P_{(x=0)}=0.295$
$P_{(x=1)}=0.235$
$P_{(x=2)}=0.164$
$P_{(x=3)}=0.1092$
$P_{(x=4)}=0.0709$
$P_{(x=5)}=0.0453$

The sum of the probabilities thus far estimated is 0.919. It would therefore be worth proceeding for two or three more values of x until the probabilities become very small. We suggest that you verify the following probabilities:

$P_{(x=6)}=0.029$
$P_{(x=7)}=0.018$
$P_{(x=8)}=0.011$

These probabilities for $x=0$ to 8 are displayed in histogram form in Fig. 7.5.

Notice the positive skew. As the value of k increases (and the variance decreases relative to the mean) so too does the degree of symmetry of the distribution, until k exceeds about 20 when the shape is nearly symmetrical (see also Fig. 8.5).

7.8 Critical probability

This chapter has demonstrated a variety of methods for generating probability values, both empirically (from samples) and fundamentally (from mathematical models). We know that all probability values lie on a scale between 0 and 1. Sometimes we wish to know the probability below which a stated outcome of an event is unlikely. Such a value is called a **critical probability** and is, of course, decided arbitrarily.

Example 7.11

A museum asks an entomologist to supply a specimen of a female *Dixella attica* to complete a reference collection. The pond referred to in Example 7.1 (where we assume the same features still apply) is visited and a collection of *Dixella* pupae obtained. How ought the entomologist to react if:

(a) the first pupa to emerge is a female *D. attica*?
(b) the first *two* pupae to emerge are females of *D. attica*?
(c) the first *three* pupae to emerge are females of *D. attica*?

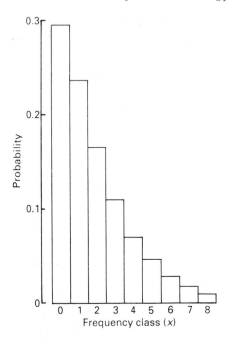

Fig. 7.5 Negative binomial probability distribution estimated from $\bar{x}=2.0$ and $k=1.33$. The height of each bar is the estimated probability that a sampling unit selected at random will contain x individuals.

From Example 7.3 we have already established that the probability that a pupa selected at random will emerge as a female *D. attica* is 0.1.

(a) If the first pupa to emerge is a female *D. attica* the entomologist should probably feel a little lucky, but there should be no cause for major celebration – this result is, after all, to be expected from 1 trial in 10.

(b) The probability that the first two emergences are *D. attica* is $0.1 \times 0.1 = 0.01$, which is to be expected from only 1 trial in 100. The entomologist should feel most fortunate that a back-up specimen has been obtained so easily. He might even harbour a suspicion that all individuals of the two insect species are not truly independent and might occur in clusters.

(c) The probability that the first three emergences are female *D. attica* is $0.1 \times 0.1 \times 0.1 = 0.001$, that is, the predicted outcome of 1 trial in 1000. If this is indeed the outcome, the entomologist has very strong grounds for considering: that the status of the insects in the pond has changed; that sampling is not random and that individuals are not independent; that pupation time for this species is shorter; that intuition is involved; or that some other unknown factor is shifting the odds favourably.

At what point does the probability of an outcome become so low that it must be regarded as 'unlikely'? Statisticians conventionally adopt three critical probability values:

1. An outcome which is predicted to occur fewer than 1 trial in 20 ($P<0.05$) is considered to be *unlikely* or *statistically significant*.
2. An outcome which is predicted to occur fewer than 1 trial in 100 ($P<0.01$) is considered to be *very unlikely* or *statistically highly significant*.
3. An outcome which is predicted to occur fewer than 1 trial in 1000 ($P<0.001$) is considered to be *extremely unlikely* or *statistically very highly significant*. The three significance levels are often indicated *, ** and ***, respectively.

Example 7.12

Consider the likelihood of the outcome $x=3$ in a Poisson distribution for which $\bar{x}=4.0$, that is to say, a sampling unit taken at random will contain three individuals.

We have estimated the probability distribution for $\bar{x}=4.0$ in Example 7.9, where $P_{(x=3)}$ was found to be 0.195. This outcome is *not unlikely* as it would be expected from nearly 1 trial in 5. The outcome is *not statistically significant*.

Example 7.13

Consider the likelihood of the outcome of $x=5$ in a negative binomial distribution for which $\bar{x}$ is 2.0 and $\hat{k}$ is 1.33.

We have estimated this probability distribution in Example 7.10, where $P_{(x=5)}$ was found to be 0.0453. This outcome is *unlikely* but not *very unlikely* – we would expect a sampling unit selected at random to contain five individuals in just under 1 trial in 20. The outcome is *statistically significant*, but *not highly significant*.

Example 7.14

In a negative binomial distribution where $\bar{x}$ is 2.0 and $\hat{k}=1.33$, how many sampling units would be expected to contain three individuals in a sample size of 150 sampling units?

From Example 7.10, the probability of $P_{(x=3)}$ for the distribution with these parameters was estimated to be 0.1092. Expected frequency = (estimated probability) × (sample size) = $0.1092 \times 150 = 16.38$. Since sampling units can only occur in whole numbers, we would expect 16 or 17 units to contain three individuals.

8 PROBABILITY DISTRIBUTIONS AS MODELS OF DISPERSION

8.1 Dispersion

The manner in which organisms are dispersed in nature is of considerable interest to field biologists. A knowledge of how individuals are dispersed is an obvious first step towards understanding their relationship with each other and their environment.

There are only three general ways in which objects may be dispersed: **regularly**, **randomly** or **contagiously** (contagious means 'aggregated' or 'clumped'). These are illustrated in Fig. 8.1. We see there is no sharp demarcation between regular and random, or between random and contagious, but rather a continuous gradient from 'crystalline' regularity to highly contagious. Often, it is perfectly obvious to which type of dispersion individuals belong. Nests of colonial birds (terns, gannets) may be spaced out in a very regular fashion, whilst eggs of many invertebrate animals (insects, molluscs) are deposited in clusters and are highly contagiously dispersed. It is not always easy to be sure, however, especially if a random dispersion is suspected.

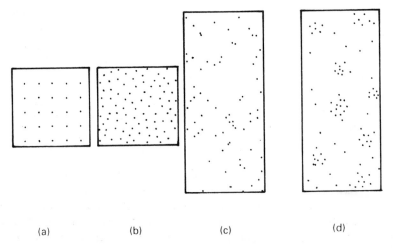

| (a) | (b) | (c) | (d) |

Fig. 8.1 Patterns of dispersion: (a) 'crystalline' regular; (b) naturally regular; (c) random; (d) contagious.

Statistical techniques can be helpful in two ways. First, they help us formulate an objective opinion as to whether a sample of count data comprises individuals which are regularly, randomly or contagiously dispersed in nature. Second, if we have obtained a sufficient number of sampling units, they allow us to determine more precisely if the distribution of counts in a sample agrees with one of the models of probability distribution. If this should be the case we can estimate the parameters of the distribution and begin to investigate underlying patterns of dispersion in nature.

It is easy to imagine *dispersion* as the way in which objects are spread out over a flat surface, such as the ground, in which case a quadrat of stated dimensions is a suitable sampling unit. But we may also wish to investigate the dispersion of events in time, in which a prescribed time interval is a suitable sampling unit. As a further example, parasites are dispersed among their hosts, and a single host is the appropriate sampling unit.

8.2 An Index of Dispersion

Let us imagine a collection of objects which are dispersed in perfect 'crystalline' regularity, as shown in Fig. 8.1(a). This would be unusual in nature but trees in an orchard or plantation might fit such a pattern. Because of the extreme regularity we would expect every sampling unit in a sample of the objects to contain exactly the same number of objects. The mean of the sample would have the same value as that of any single observation, and the standard deviation – and hence the variance – of the sample would be zero. In 'naturally' regular dispersion we would expect a little variation between the observations in a sample, but the variance would still be low.

Turning to the other extreme of the dispersion gradient, samples of objects which are dispersed contagiously will contain some sampling units with low counts (0, 1) but with the occasional unit with a very high count. The value of the mean of such a skewed distribution may not be obvious before calculation but the standard deviation, and variance, will be large. Between these extremes, samples of objects which are dispersed randomly have intermediate variances. We can now make the following tentative generalization:

- a sample of count data with a small variance suggests *regular dispersion*
- a sample of count data with intermediate variance suggests *random dispersion*
- a sample of count data with a large variance suggests *contagious dispersion*

We need a quantitative scale to demarcate the sections of the dispersion gradient (regular, random, contagious). In Section 7.6 we note that the Poisson distribution is the model which best describes a random dispersion of objects, Furthermore, it can be shown mathematically that in a Poisson distribution the variance of σ^2 equals the mean μ. Therefore, in a sample of randomly dispersed objects, we expect the sample variance s^2 to approximately equal the sample mean $\bar{x}$. That is, the ratio of the variance to the mean $s^2/\bar{x} \simeq 1$. This ratio

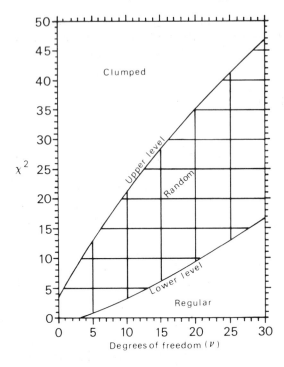

Fig. 8.2　95% confidence zone of random dispersal.

may be used as an *Index of Dispersion* with the value unity representing random dispersion. Our generalization can be made more precisely:

- objects dispersed regularly yield samples where $s^2/\bar{x} < 1$
- objects dispersed randomly yield samples where $s^2/\bar{x} \simeq 1$
- objects dispersed contagiously yield samples where $s^2/\bar{x} > 1$

Finally, we require an objective way to find the critical values of the ratio which separate regular from random, and random from contagious. To do this, the variance/mean ratio is *standardized* by multiplying by the number of observations in the sample minus one $(n-1)$, that is, the degrees of freedom v. We give this product the symbol χ^2 (chi-square). Consult the graph in Fig. 8.2. Find the point which corresponds to the coordinates of the calculated χ^2 and the degrees of freedom $(n-1)$. The tables from which the graph is constructed inform us that the conclusion will be reliable on 95 times out of every 100 that the graph is used.

Example 8.1

As part of a marine oil pollution monitoring programme near an oil refinery, biologists investigate annually the density of limpets (*Patella* sp.) inhabiting suitable slabs of bedrock in the intertidal zone near the refinery. The sampling unit is a quadrat of 20×20 cm (0.04 m^2) and a sample comprises observations from 20 sampling units. A sample contains the following observations:

11 6 8 9 9 10 7 8 7 7 9 8 10 7 8 8 9 11 9 8

How are the limpets dispersed over the rock face?

The steps in the procedure are:

 (i) Use a scientific calculator to obtain the mean, $\bar{x}$, and the standard deviation, s. These are 8.45 and 1.356, respectively.
 (ii) Square the standard deviation to obtain the variance, s^2. This is 1.84.
 (iii) Calculate the Dispersion Index by dividing the variance by the mean. This is $1.84 \div 8.45 = 0.218$. The value is well below 1 and a regular dispersion may already be suspected.
 (iv) Calculate the factor χ^2 by multiplying the Dispersion Index by the degrees of freedom v from $(n-1)=(20-1)=19$. The result is 4.14.
 (v) Find the point on the graph in Fig. 8.2 at the intersection of the values of χ^2 and v (4.14 and 19).
 (vi) The point is well below the lower line demarcating the random dispersion, and a regular dispersion is accepted.

The investigator may now wish to try to explain *why* the limpets are dispersed regularly and may study factors such as intra-specific competition for grazing space on the rock surface.

Example 8.2

A biologist studying pollination observes the frequency of visits by bees to a flower bed. The sampling unit is a 1-min interval and in 10 intervals the following counts are obtained:

4   6   3   3   9   4   6   2   5   2

How are the visits by bees dispersed?

Following the procedure outlined in Example 8.1, we determine that:

$\bar{x}=4.4$; $s=2.17$; $s^2=4.71$

The Dispersal Index $s^2/\bar{x}$ is 1.07 which is close to 1. The intersection of χ^2 ($1.07 \times 9 = 9.63$) and v ($10-1=9$) in Fig. 8.2 lies within the random zone. A random dispersion is therefore accepted.

Example 8.3

A biologist investigates the dispersion of the green-winged orchid (*Orchis morio*) in a hay-meadow nature reserve. Twelve 1×1 m quadrats (sampling units) are placed at random points in the meadow and the number of vegetative basal rosettes are counted. The following counts are obtained:

0   0   4   2   58   1   0   22   5   7   17   1

How is the orchid dispersed?

Following the procedure outlined in Examples 8.1 and 8.2 we determine that the Dispersion Index $s^2/\bar{x} = 281/9.75 = 28.8$. The value is much greater than 1 and a contagious dispersion is suspected. The intersection of χ^2 ($28.8 \times 11 = 316.8$) and v ($12 - 1 = 11$) on the graph in Fig. 8.2 is clearly way above the boundary of the random zone and a contagious dispersion is accepted. With such a high degree of contagiousness, there are strong grounds for doubting that individuals are independent; perhaps they are connected by rhizomes, or can only grow in localized areas where conditions are perfect.

8.3   Choosing a model of dispersion

In Section 8.2 we show how small samples (< 30 sampling units) of count data may be classified as having been drawn from regularly, randomly or contagiously dispersed populations. A key step is the calculation of a Dispersion Index from the ratio of the variance of a sample to its mean.

When a sample has more than about 30 observations, the observations can be grouped into a frequency distribution. It is then possible to find out if the shape of the distribution agrees well with the shape that would be expected from one of the common probability distributions. Where this is the case, there is then a very high degree of confidence about the nature of the dispersion. There are two problems: first, a suitable model must be chosen (that is, it must be statistically 'reasonable'); second, the expected distribution has to be worked out. The guidelines for choosing a model are:

- if the Dispersion Index suggests *regular dispersion*, choose *binomial*
- if the Dispersion Index suggests *random dispersion*, choose *Poisson*
- if the Dispersion Index suggests *contagious dispersion*, choose *negative binomial*.

The general procedure for working out the expected distribution is the same for each model:

(i) Estimate the parameters of the distribution from the sample data.

(ii) Use the parameter estimates to estimate the probability distribution over a suitable range of frequency classes (often this will be 0, 1, 2 and so on until the highest frequency class in the sample).

(iii) Convert the expected *probability* distribution into an expected *frequency* distribution by multiplying the probability of each frequency by the sample size.

(iv) Superimpose the *expected* frequencies over the *observed* frequencies to see how well they agree – in other words to see if there is a 'good fit'.

We now describe the details of the procedure for each model in turn.

8.4 The binomial model

Computation of the expected probabilities of a binomial distribution is based on the procedure described in Example 7.6. A word of explanation about the estimation of the parameters k and p is required. In Example 7.6 the value of k was decided arbitrarily as a whole number of events or trials. In using the binomial distribution as a model of regular dispersion, the parameter k is equivalent to the highest number of individuals that could be contained in a sampling unit. As we are often unsure of what this is, k is estimated (rounded to a whole number) from the sample data thus:

$n = $ sample size.

$$\hat{k} = \frac{\bar{x}^2}{(\bar{x} - s^2)}$$

In Example 7.6, the parameter p is the probability of one of the two possible outcomes of a single event and q is the probability of the other. In a regular dispersion, p is taken to be the probability that any point in a sampling unit is occupied by an individual. An estimate of p is obtained by:

$$p = \frac{\bar{x}}{k} \text{ and therefore } q = (1 - p)$$

Example 8.4

Referring to the limpet density study outlined in Example 8.1 a more detailed investigation of the dispersion of limpets was called for. A frequency table of limpets counted in 50 sampling units (20×20) cm quadrats) is given below:

Number of limpets x:	6	7	8	9	10	11
Frequency f:	4	6	11	16	9	4

Thus, four quadrats contained 6 limpets, six quadrats contained 7 limpets, and so on. The sample data are calculated by means of a scientific calculator:

$n = 50$; $\bar{x} = 8.64$; $s = 1.352$; $s^2 = 1.828$

The variance is much less than the mean, so a binomial model is appropriate. From the sample statistics we estimate:

$$\hat{k} = \frac{8.64^2}{(8.64 - 1.828)} = 10.96 \text{ which is 11 rounded to a whole number}$$

$$p = \frac{8.64}{11} = 0.785$$

$$q = (1 - p) = 0.215$$

Using the binomial equation given in Section 7.5 we can show that for values of x between 0 and 4 the probability is so small it can be ignored. We start at $x = 5$.

For $P_{(x = 5)}$

$$P_{(x=5)} = \frac{k!}{x!(k-x)!} \times p^x \times q^{(k-x)}$$

$$= \frac{11!}{5!(6!)} \times 0.785^5 \times 0.215^6$$

$$= 0.0136$$

Therefore, $f'_{(x=5)}$ (the expected frequency for $x=5$) is $P_{(x=5)} \times n$, that is, $0.0136 \times 50 = 0.68$.

For $P_{(x=6)}$

Similarly, $P_{(x=6)} = \dfrac{11!}{6!(5!)} \times 0.785^6 \times 0.215^5 = 0.050$

and

$$f'_{(x=6)} = P_{(x=6)} \times n = 2.5$$

In like manner we calculate:

$P_{(x=7)} = 0.129$	$f'_{(x=7)} = 6.45$
$P_{(x=8)} = 0.236$	$f'_{(x=8)} = 11.8$
$P_{(x=9)} = 0.288$	$f'_{(x=9)} = 14.4$
$P_{(x=10)} = 0.211$	$f'_{(x=10)} = 10.55$
$P_{(x=11)} = 0.07$	$f'_{(x=11)} = 3.5$
$P_{(x=12)} = 0.021$	$f'_{(x=12)} = 1.05$

   We are now in a position to compare the *observed* frequency distribution with the *expected* distribution. Even though the expected frequencies are not whole numbers, they are *not* rounded for the purposes of comparison. In Fig. 8.3 we show the observed frequencies as a histogram and the expected frequencies superimposed as joined closed circles. The agreement is clearly very good – the small discrepancies between the observed and expected frequencies are easily accounted for by random scatter inherent in any sample.

8.5 Poisson model

The Baermann apparatus is a device for extracting microscopic nematode worms from soil cores. A core of soil (about $50\,\text{cm}^3$) is dispersed in water and suspended over a coarse filter supported in a funnel. Nematodes gravitate through the filter down into the stem of the funnel where they accumulate above a tap. After a suitable extraction interval (about 24 h) a volume of water is run through the tap into a counting chamber mounted on a microscope slide. The base of the chamber has an arbitrary but regular grid of 6×10 squares etched into it (rather like a haemocytometer). The slide is tracked under a microscope and the number of nematodes in each square counted and recorded. Nematodes appear to be randomly dispersed over the base of the

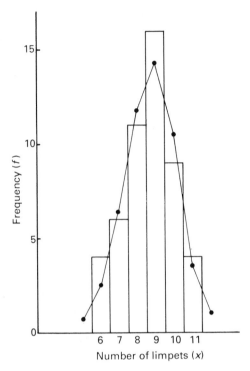

Fig. 8.3 Frequency distribution of limpets in 50 quadrats. Histogram is observed frequency and joined closed circles are frequencies expected from a binomial model.

chamber and there is no obvious attraction to, or repulsion from, each other or the sides of the counting chamber.

Example 8.5

A frequency table of the numbers of nematodes counted in all 60 squares of a counting chamber is presented below. Does the dispersion of worms over the base of the chamber constitute a Poisson distribution?

Number of nematodes, x:	0	1	2	3	4	5	6	7	8
Frequency, f:	3	12	17	13	9	3	1	2	0

Thus, three squares contain 0 nematodes, 12 squares contain 1, and so on.

The sample statistics are calculated on a scientific calculator:

$$n = 60; \quad \bar{x} = 2.6; \quad s = 1.564; \quad s^2 = 2.446$$

The variance is similar to the mean and the Poisson model is appropriate. The Poisson distribution is determined by a single parameter, λ, which is estimated from a sample by the mean, $\bar{x}$. Following the procedure described in

Example 7.9, we estimate the expected probability for each frequency class x by:

$$P_{(x)} = e^{-\bar{x}} \times \frac{\bar{x}^x}{x!}$$

For $P_{(x=0)}$

$$P_{(x=0)} = 2.7183^{-2.6} \times \frac{2.6^0}{0!} = 0.0743$$

Therefore, $f'_{(x=0)} = (P_{(x=0)}) \times n = 0.0743 \times 60 = 4.458$

For $P_{(x=1)}$

$$P_{(x=1)} = 2.7183^{-2.6} \times \frac{2.6^1}{1!} = 0.193$$

and,

$$f'_{(x=1)} = 0.193 \times 60 = 11.58$$

For $P_{(x=2)}$

$$P_{(x=2)} = 2.7183^{-2.6} \times \frac{2.6^2}{2!} = 0.251$$

and

$$f'_{(x=2)} = 0.251 \times 60 = 15.06$$

In similar manner we calculate:

$P_{(x=3)} = 0.218$ $f'_{(x=3)} = 13.08$
$P_{(x=4)} = 0.141$ $f'_{(x=4)} = 8.46$
$P_{(x=5)} = 0.074$ $f'_{(x=5)} = 4.44$
$P_{(x=6)} = 0.032$ $f'_{(x=6)} = 1.92$
$P_{(x=7)} = 0.012$ $f'_{(x=7)} = 0.72$
$P_{(x=8)} = 0.00385$ $f'_{(x=8)} = 0.231$

The observed frequencies are shown as a histogram in Fig. 8.4 with the expected frequencies superimposed. The agreement is very good, the small discrepancies being easily accounted for by random scatter. The dispersion of the nematodes is adequately described in terms of a Poisson model.

8.6   The negative binomial model

In Section 7.7 we said that the negative binomial is a robust model which describes a wide range of natural situations in which objects are clumped. The dispersion of parasites among their hosts often shows a very good agreement with the distribution predicted by a negative binomial model.

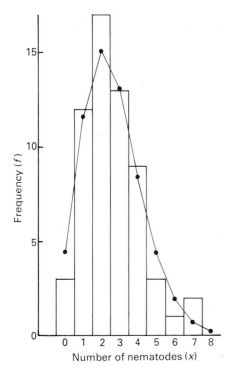

Fig. 8.4 Frequency distribution of nematodes in 60 grid squares of a counting chamber. Histogram is observed frequency and joined closed circles are frequencies expected from a Poisson model.

The bank vole *Clethrionomys glareolus* is a popular subject in small mammal studies; it is easily captured in small traps and is easy to handle. During routine examination fleas can be found by stroking back the fur.

Example 8.6

During a study of a population of bank voles 80 animals were captured and examined for fleas. Upon capture each vole was marked to prevent repeat observations if recaptured, thus ensuring that each sampling unit – a vole – was independent. A frequency table showing the distribution of fleas on the voles is given below:

Number of fleas x:	0	1	2	3	4	5	6	7	8
Frequency f:	18	25	15	10	7	2	2	1	0

Thus, 18 voles had no fleas, 25 had 1 flea, and so on. Can this distribution be described in terms of a negative binomial model?

The sample statistics are calculated by means of a scientific calculator:

$n = 80$: $\bar{x} = 1.78$; $s = 1.61$; $s^2 = 2.59$

The variance is greater than the mean and so a negative binomial model is appropriate.

From the sample statistics we estimate the parameters of the distribution (see Section 7.7). $\bar{x}$ is a satisfactory estimate of μ, namely 1.78; $\hat{k}$ is estimated from:

$$\hat{k}=\frac{\bar{x}^2}{(s^2-\bar{x})}=\frac{1.78^2}{(2.59-1.78)}=3.91$$

Adopting the procedure described in Example 7.10, we estimate the probability for each frequency class x, starting at $x=0$.

For $P_{(x=0)}$

$$P_{(x=0)}=\left(1+\frac{\bar{x}}{k}\right)^{-\hat{k}}=\left(1=\frac{1.78}{3.91}\right)^{-3.91}$$

$$=0.231 \qquad \text{[Remember to store this in your calculator's memory!]}$$

Therefore, $f'_{(x=0)}=P_{(x=0)}\times n=0.231\times 80=18.48$

Calculate the factor **F** from $\dfrac{\bar{x}}{\bar{x}+\hat{k}}=\dfrac{1.78}{1.78+3.91}=0.313$

Enter this number as a constant in your calculator, or write it down.

For $P_{(x=1)}$

$$P_{(x=1)}=\frac{\hat{k}}{1}\times \mathbf{F}\times P_{(x=0)}$$

$$=3.91\times 0.313\times 0.231 \qquad \text{[Recall 0.231 from memory then clear memory]}$$
$$=0.283 \qquad \text{[Store 0.283 in memory]}$$

Therefore, $f'_{(x=1)}=0.283\times 80=22.64$.

For $P_{(x=2)}$

Similarly, $P_{(x=2)}=\dfrac{(3.91+1)}{2}\times 0.313\times 0.283 \qquad \text{[Recall 0.283 from memory then clear memory]}$

$$=0.217 \qquad \text{[Store in memory]}$$

Therefore, $f'_{(x=2)}=0.217\times 80=17.36$

In similar manner we calculate:

$P_{(x=3)}=0.134 \qquad f'_{(x=3)}=10.72$
$P_{(x=4)}=0.072 \qquad f'_{(x=4)}=5.76$
$P_{(x=5)}=0.036 \qquad f'_{(x=5)}=2.88$
$P_{(x=6)}=0.017 \qquad f'_{(x=6)}=1.36$
$P_{(x=7)}=0.008 \qquad f'_{(x=7)}=0.64$
$P_{(x=8)}=0.003 \qquad f'_{(x=8)}=0.24$

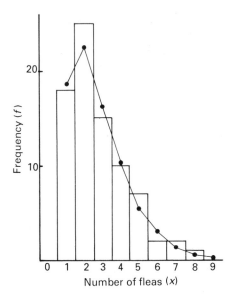

Fig. 8.5 Frequency distribution of fleas on 60 bank voles. Histogram is observed frequency and joined closed circles are frequencies expected from a negative binomial model.

The observed frequencies are shown as a histogram in Fig. 8.5, with the expected negative binomial frequencies superimposed. There is clearly an excellent agreement between the observed and the expected frequencies. The dispersal of fleas on bank voles is adequately described in terms of a negative binomial model.

8.7 Deciding the goodness of fit

For illustrative purposes we have selected an example of each probability distribution described – binomial, Poisson and negative binomial – in which there is close and undoubted agreement between the observed and expected frequencies. We have attributed small discrepancies between the observed and expected frequencies to *chance scatter* which is inherent in any sample that purports to estimate population parameters. In cases where the discrepancies are considerable, we may doubt that the observed data fit those predicted by the chosen model and are obliged to reject any notion that the data can be described in terms of that model. The point at which the scatter is so large that agreement with a model cannot be accepted is decided by a statistical test. 'Goodness of fit' tests are the subject of Section 13.6.

9 THE NORMAL DISTRIBUTION

9.1 The normal curve

In Fig. 4.7 we showed the distribution of 100 shoot lengths measured to a precision of ± 0.5 mm. Because measurements of length are on a continuous scale, they may be made with increased precision to, say, ± 0.05 mm. Each frequency class then corresponds to increments of 0.1 mm and, if the number of observations is accordingly increased, the steps in the histogram become more gradual as shown in Fig. 9.1(a). As increasing numbers of observations are obtained with increasing degrees of precision, the histogram verges towards a smooth, symmetrical bell-shaped curve as shown in Fig. 9.1(b).

The mathematician Gauss discovered that distributions of this shape estimated from large samples of measurements drawn from a single population often agree well with a model of the following form:

$$y = \frac{1}{\sigma\sqrt{(2\pi)}}\, e^{-[(x-\mu)^2/2\sigma^2]}$$

This is the equation of the **normal curve**, and is revealing in several ways. It allows us to compute the height of the curve y for a nominated value, x. Before we can do this we must assign values to the other symbols in the equation. The terms π and e are, of course, mathematical constants (equal to 3.14 and 2.72, respectively). The remaining terms are μ (the population mean) and σ (the population standard deviation). These are the parameters of the distribution. Usually we do not know the values of μ and σ and have to estimate them from a

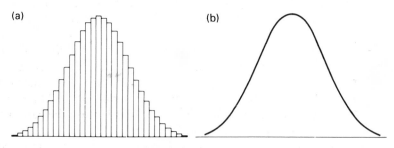

Fig. 9.1 Gradation of a histogram (a) into the normal curve (b).

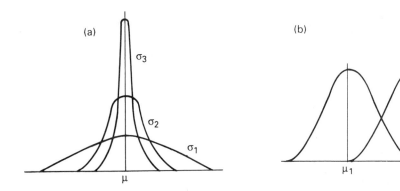

Fig. 9.2 Normal curves. (a) Similar μ and different σ; (b) different μ and similar σ.

sample as $\bar{x}$ and s. If the number of observations in the sample exceeds about 30, $\bar{x}$ and s are reliable estimates of the parameters.

As a continuous variable, x may assume any degree of precision and therefore y is also continuous – producing a smooth, continuous distribution curve. This contrasts with the three distributions described in Chapter 8 in which values of x are whole numbers. We observe also that as the term $(x - \mu)$ becomes smaller, y becomes larger and is at a maximum when $x = \mu$. This results in a symmetrical curve in which the apex occurs where x is equal to the population mean. These features are shown in Fig. 9.2. Figure 9.2(a) shows curves with similar μ and different σ; Fig. 9.2(b) shows two curves of different μ and similar σ. In reality, these parameters may vary continuously generating an infinite variety of normal curves.

9.2 Some mathematical properties of the normal curve

A normal curve is symmetrical, with the axis of symmetry passing through the baseline where $x = \mu$, that is, through one of the parameters of the curve. Theoretically, the two tails of the curve never actually touch the horizontal axis; rather, they continue to approach it over an infinite distance. In real life, however, the tails are never very long – we do not come across members of a single population that are both astronomically large and microscopically small.

On each side of the curve is a point of inflection where the direction of the curve changes from concave to convex. It so happens that if a vertical line is dropped down from the point of inflection it cuts the baseline at a distance on either side of the central axis equal to the second parameter, σ, the standard deviation. This distance can be used as a standard unit to divide up the baseline (the x-axis) into equal segments, as shown in Fig. 9.3.

If the vertical axis of the distribution is re-scaled by dividing by the number of observations, it effectively becomes a probability distribution or, strictly, a

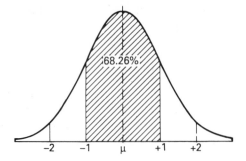

Fig. 9.3   Normal curve with standard deviations.

probability density. The total probability encompassed by the density is 1. If we say that the total *area* under the curve is 100% then one of the mathematical properties of the normal curve is that the area bounded by one standard deviation on each side of the central axis (that is, $\mu \pm \sigma$) is approximately 68.26% of the total area. This means that about 68% of observations drawn randomly from a normally distributed population will fall within ± 1 standard deviation from the mean. The other 32% fall outside these limits, 16% above 1 standard deviation and 16% below 1 standard deviation. In other words, there is a probability of $P = 0.68$ that a single observation drawn at random will fall within $\mu \pm \sigma$. By extending the limits to *two* standard deviations ($\mu \pm 2\sigma$) the proportion of observations that will be included within them is increased to 95.44%. Similarly, 99.74% of observations fall within $\mu \pm 3\sigma$. These are equivalent to probabilities of 0.9544 and 0.9974. In practice the values of 0.95 and 0.99 are more convenient to deal with. It can be calculated that these probability values fall at $\mu \pm 1.96\sigma$ and $\mu \pm 2.58\sigma$, respectively. It follows that the probability of a random observation falling *outside* these limits is 0.05 and 0.01. It will be recalled from Section 7.8 that these are the critical probability values for assessing whether the outcome of a stated event is *unlikely* or *very unlikely*. Therefore, we may use the properties of the normal curve to assess the likelihood of certain outcomes, as we show in Section 9.3.

9.3   Standardizing the normal curve

Any value of an observation x on the baseline of a normal curve can be standardized as a number of **standard deviation units** the observation is away from the population mean, μ. This expression is called a **z-score**. To transform x into z apply the formula:

$$z = \frac{(x - \mu)}{\sigma}$$

If μ is larger than x, then z is negative.

It is often the case that we do not know the values of μ and σ. In samples of more than about 30 observations, z is given by:

$$z = \frac{(x - \bar{x})}{s}$$

By **standardizing** an observation x into a z-score, we can relate it to the properties which apply to all normal curves. Thus, if the calculated value of z is larger than 1.96, then the probability of such an observation drawn at random is less than 0.05. It is regarded as *unlikely* or **statistically significant.**

Example 9.1

On the basis of very large samples the mean (μ) length of a population of seeds is estimated to be 3.8 mm and σ is estimated to be 0.15 mm. Is it likely that a randomly selected seed of length 4.3 mm belongs to this population?

Convert the observation x to a z-score:

$$z = \frac{(x - \mu)}{\sigma} = \frac{(4.3 - 3.8)}{0.15} = 3.33$$

The calculated value of z is much larger than 1.96 which is equivalent to $P = 0.05$. Indeed, it is larger than 2.58 which is equivalent to $P = 0.01$. We say that the observation is **statistically highly significant** and conclude that it is very unlikely that the individual seed belongs to the same population.

This computation and our resulting conclusion is an example of a simple statistical test. We will recall it when we outline the principles of statistical testing in Chapter 12.

9.4   Two-tailed or one-tailed?

In Section 9.2 we said that under a normal curve 95% of observations are contained by $\mu \pm 1.96\sigma$, and that the residual 5% is divided equally outside these limits in both tails of the distribution. This is shown in Fig. 9.4.

There is an alternative way of apportioning 95% of the observations and that is by excluding the residual 5% in a single tail (either the *upper* tail or the *lower* tail). In this case the cut-off point occurs at 1.65 standard deviations on whichever side of the distribution the 5% is excluded. This is shown in Fig. 9.5.

Sometimes we are able to say that if a random observation probably does *not* belong to a population of specified mean and standard deviation, then it *must* belong to a population that has a mean which is specified as being larger or, alternatively, to a population that has a mean which is specified as being smaller. In this case, the critical value of z is set at the lower value of 1.65 and the test is said to be *one-tailed.*

Reference to Example 9.1 may help to make this clear. Before the seed is

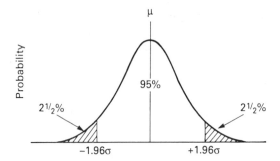

Fig. 9.4 The normal curve with 5% in two tails.

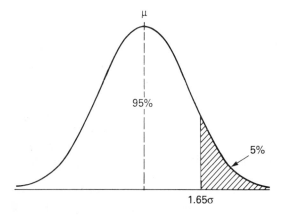

Fig. 9.5 The normal curve with 5% in upper tail.

measured the question is posed, 'Is its size consistent with a population of specified mean and standard deviation?' If the test shows that it is *not* consistent, then there is an equal chance that it could come from a population whose mean is either larger or smaller. The test is therefore *two-tailed* and the critical value of z is set at 1.96. Consider the following example.

Example 9.2

The mean length of a male insect is estimated from large samples to be 5.6 mm and the standard deviation 0.26 mm. Females of the same species are known to be larger. An individual of length 6.1 mm is found. Is it likely to be an example of a male?

In this case, if we can show that an observation of this magnitude is unlikely to have come from a normally distributed population with parameters of the stated values, then it *must* have come from one with a larger mean, that is, a female. The test is therefore *one-tailed* and we set the critical value of z at 1.65.

chi-square always 1 tailed

Convert the individual observation x to a z-score:

$$z = \frac{(x - \mu)}{\sigma} = \frac{(6.1 - 5.6)}{0.26} = 1.9$$

The computed value of z is 1.9. This is smaller than the critical value of z of 1.96 in a two-tailed test, but larger than that of 1.65 in a one-tailed test. Since the test is one-tailed, the result is *statistically significant*, and we conclude that it is unlikely that the individual is a male. It is probably a female.

A word of warning: The critical value of z in the one-tailed test is considerably lower than that in the two-tailed test. Often it is not possible to be sure that a test is genuinely one-tailed. If you are not sure, always assume that it is two-tailed and set the critical value of z at 1.96 (or higher, e.g. 2.58, if greater stringency is required).

9.5 Small samples: the *t*-distribution

In the normal distribution the z-scores of 1.96 and 2.58 indicate the limits on either side of a population mean within which 95% and 99% of all observations will fall. Usually we do not know the value of μ and σ and are obliged to estimate them from a sample. We can only be confident that a sample mean and standard deviation are reliable estimates of the population parameters if the sample is large. When the sample is small, we are less confident. To compensate for the uncertainty the values 1.96 and 2.58 are increased, that is, set further out from the mean and the symbol z is replaced by the symbol t. As the sample size decreases, so too does the degree of certainty. The values of t must therefore increase accordingly. Thus, in addition to the mean and standard deviation parameters, t-distributions are also dependent on sample size. However, it is not n which determines t directly, but $(n - 1)$, that is, the degrees of freedom. There are mathematical techniques available for working out the complete probability distributions of t for any number of degrees of freedom. In practice we usually need to know only the values which correspond to $P = 0.05$ and $P = 0.01$ for a particular sample size, that is, the values which correspond to the z-score of 1.96 and 2.58 in large samples. These values are found in statistical tables.

In the same way that an observation can be converted to a z-score if the estimates μ and σ are available from a large sample, so too, an observation x may be transformed to a t-score when the values of the mean and standard deviation of a small sample are known. Thus:

$$t = \frac{(x - \bar{x})}{s}$$

If the calculated value of t is larger than that tabulated for $(n - 1)$ degrees of freedom at $P = 0.05$, it is concluded that the observation is unlikely to have been drawn from a population with the same mean as that from which the sample is drawn.

In the examples that follow we refer to the tables for t in Appendix 2.

Example 9.3

How many standard deviations on either side of a sample mean derived from 10 observations would be expected to contain (i) 95% and (ii) 99% of the observations of the normally distributed population from which the sample is obtained?

(i) If the sample comprises 10 observations there are 9 degrees of freedom (df). Enter the table in Appendix 2 at 9 df and run along until the column headed 0.05 (two-tailed test) is reached. The value is 2.262. Therefore 95% of the observations in the population fall within 2.262 standard deviations on either side of the sample mean, that is, $\bar{x} \pm 2.262s$.
(ii) Similarly, the value of t for 99% is found in the column headed 0.01 against 9 df, that is 3.250.

Example 9.4

A sample of 12 spikelets of a species of meadow grass (*Poa* sp.) is obtained randomly from different flower heads. The lengths of the spikelets are (mm):

4.5   4.9   6.6   5.3   5.2   6.1   5.4   6.2   5.2   4.7   5.6   5.5

A single meadow grass spikelet found loose in a herbarium measures 4.4 mm. Is it likely that this specimen represents a population with the same mean as that from which the sample is drawn?

Using a scientific calculator, we determine that the mean and standard deviation of the sample are 5.43 mm and 0.62 mm, respectively. Convert the observation of the single spikelet x into a t-score:

$$t = \frac{(x - \bar{x})}{s} = \frac{(4.4 - 5.43)}{0.62} = -1.66$$

Because we are using two-tailed tables we ignore the minus sign. Enter the table in Appendix 2 at $(12 - 1) = 11$ df. Under the heading 0.05 we find the value 2.201; our calculated value of 1.66 if less than this. We conclude that a spikelet of length 4.4 mm is consistent with the population from which the sample is drawn. Of course this does not prove that the spikelet comes from the same population; rather, it fails to provide evidence that it does not.

Examination of the t-table for $P = 0.05$ shows that as the degrees of freedom increase the value of t decreases until it verges towards the value of z (1.96) when $n = \infty$ (infinity). Above 30 df, however, the relative change in the value of t becomes very small with increasing sample size; it is very slightly larger than 2.0. The value of z at $P = 0.05$ is only slightly smaller than 2.0. Thus, when the sample size exceeds about 30, the difference between z and t may often be ignored. In published accounts of their work many biologists often settle for the arithmetic convenience of $\bar{x} \pm 2$ standard deviations as the critical limit unless samples are very small.

9.6 Are our data 'normal'? *[handwritten: ARE THEY NORMALLY DISTRIB]*

Many parametric statistical techniques that we describe depend on the mathematical properties of the normal curve. They usually assume that samples have been drawn from populations which are normally distributed. Sometimes, if samples are very small, it is hard to know if the parent population is normal, in which case **distribution free (non-parametric)** techniques are appropriate. Often a simple check will reveal a serious departure from normality. Before we show you this procedure we consider some types of samples that can never be regarded as having been drawn from normally distributed populations.

Because the normal curve is continuous, samples which comprise units of *count* data can never in theory, be *normal*: they are more likely to conform to a binomial, Poisson or negative binomial distribution. However, the error due to the lack of continuity, that is, due to the 'steps' in a frequency distribution, need not be serious if the distribution is fairly symmetrical. We have seen that this is the case in a binomial distribution when P is close to 0.5 and in a Poisson distribution when $\bar{x}$ is larger than about 5.0. There we may regard the distribution as being *approximately normal*.

Errors are more likely to be serious when data are strongly skewed such as we invariably find in the negative binomial distribution or in any distribution of count data in which the variance is greater than the mean. Data of these kind should not be used in parametric tests without first adjusting or *transforming* the data to *normalize* them. Chapter 10 considers methods of transforming count data.

Even samples of measurement data may depart seriously from normality under some circumstances. Many organisms are polymorphic, for example sexually dimorphic. Samples of measurements of individuals from such populations will be multi-modal with each mode probably corresponding to a population class, for example sex or age-class. Before any sensible statistical analysis can be undertaken the individual classes should be separated out and analyses conducted on each class individually (see Section 5.4).

How can we check if samples are likely to have been drawn from a normally distributed population? It is possible to apply the Gaussian equation (Section 9.1) to estimate expected frequencies from the sample statistics and then undertake a *goodness-of-fit* test (see Section 8.7). This however is tedious and usually unnecessary. There is a simpler alternative: plot out the data and see if they *look* normal! As a back-up, calculate the mean and standard deviation of *[handwritten: NB]* the sample and see if about 70% of the observations fall within the interval $\bar{x} \pm s$.

[handwritten: 'Checking if observations are normally distributed.]

Example 9.5

Three different samples with 10 observations in each are shown below. Do they appear to have been drawn from normally distributed populations?

(a)

x	f
6	1
7	1
8	3
9	2
10	2
11	1

(b)

x	f
5	3
6	3
7	2
8	1
9	1

(c)

x	f
5	3
6	2
7	1
8	2
9	2

(*a*) *A dot-diagram shows*:

```
            .
        .   .   .
    .   .   .   .   .       .
  _____
  6   7   8   9   10   11
```

$\bar{x} = 8.6$; $s = 1.5$
$\bar{x} \pm s$ is 7.1 to 10.1 which contains 7/10 observations

Although the sample is slightly skewed, the observations are scattered on either side of a mode and 7 out of the 10 observations fall within $\bar{x} \pm s$. There is no reason to doubt that the sample is drawn from a normal population.

(*b*) *A dot-diagram shows*:

```
  .   .
  .   .   .
  .   .   .   .   .
  _____
  5   6   7   8   9
```

$\bar{x} = 6.4$; $s = 1.35$
$\bar{x} \pm s$ is about 5.05 to 7.75 which contains 5/10 observations

In this case there is a perceptible skew in the data. Only half the observations fall within $\bar{x} \pm s$. We doubt that the sample is drawn from a normally distributed population.

(*c*) *A dot-diagram shows*:

```
  .
  .   .       .   .
  .   .   .   .   .
  _____
  5   6   7   8   9
```

$\bar{x} = 6.8$; $s = 1.62$
$\bar{x} \pm s$ is 5.18 to 8.42 which contains 5/10 observations

In this case there is marked bimodalism. $\bar{x} \pm s$ contains only half the observations. We doubt that the sample is drawn from a normally distributed population.

0 DATA TRANSFORMATION

0.1 The need for transformation

In Chapter 9 we explained how an infinite variety of normal curves may be generated by varying the values of the two parameters μ and σ. In every one of such curves a constant proportion of its area is enclosed by $\mu \pm$ a given number of standard deviation units. A number of important statistical techniques depend upon this property and their correct application assumes that samples have been drawn from populations which are normally distributed. Many samples of 'count' data however are drawn from populations which are strongly skewed; they are more likely to represent binomial, Poisson or negative binomial distributions. Under some circumstances when the values of the respective parameters describe fairly symmetrical distributions (for example, in a binomial distribution when p is near to 0.5) then the distributions are 'approximately normal'. When a distinct skew exists the techniques may only be applied without risk of error if the data are first *normalized* by transformation.

There is a second major reason why transformation may be necessary. Parametric statistical techniques which compare the means of two or more samples assume that the variances of each sample are so similar that differences between them may be ignored. Samples of observations which are counts of things are unlikely to meet this condition. The reason why is most easily understood in the case of the Poisson distribution. This distribution is determined by a single parameter λ which is equal to the mean and variance of a $\lambda = m$ population. It follows, therefore, that populations with large means will have larger variances; in other words the variance of a population is dependent upon the mean. This relationship is reflected in samples drawn from the populations. A similar association is found in binomial and negative binomial distributions. In both models, larger means result in more symmetrical distributions in which a greater spread of observations on both sides of the mean is possible. The greater spread, of course, results in a greater variance (Fig. 10.1).

It is a fortuitous mathematical fact that transformation techniques which normalize data also tend to remove the dependency of the variance upon the mean. Transformation is said to *stabilize the variance*. Transformation means simply the conversion of the raw values of all observations, x, into a mathematical derivative. The three most widely applied in biology are logarithmic, square root and arcsine transformations. Each is used in different circumstances. We now describe the application of each in turn.

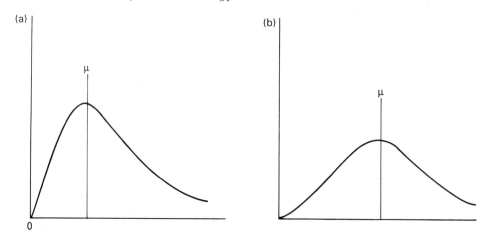

Fig. 10.1   (a) Small μ: spread of observations below μ is truncated by zero. (b) Larger μ: spread of observations below μ is less restricted.

10.2   The logarithmic transformation

The logarithmic transformation is appropriate when the variance of a sample of count data is larger than the mean (see Examples 8.3 and 8.6). In this transformation, each observation x is replaced by the log of itself: that is, replace x by $\log x$. The effect of the transformation is demonstrated by the data in Table 10.1. We may imagine they refer to the numbers of a particular item counted in 88 sampling units. Thus, 2 units have three items, 4 units have four items, and so on. Figure 10.2(a) shows a frequency bar graph of the raw data. Notice the positive skew. The distribution is clearly far from being normal. In Fig. 10.2(b) the distribution is shown with the horizontal x-axis re-scaled (*transformed*) by replacing each value of x (3,4,5 . . . 20) by its logarithm. Notice how the right-hand tail has been squashed up, making the overall shape more symmetrical – much more like the shape we expect to see in a normal distribution. The transformed shape, of course, can never be *exactly* normal (remember that a normal curve is continuous, and we are dealing here with

Table 10.1   Logarithmic transformation of 88 observations

x:	3	4	5	6	7	8	9	10	11
log x:	0.477	0.602	0.699	0.778	0.845	0.903	0.954	1.00	1.04
frequency f:	2	4	6	8	11	11	10	8	6

x:	12	13	14	15	16	17	18	19	20
log x:	1.08	1.11	1.15	1.18	1.20	1.23	1.26	1.28	1.30
frequency f:	5	4	3	3	2	2	1	1	1

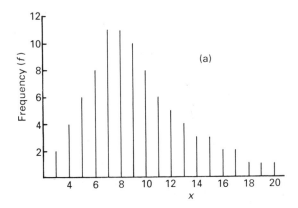

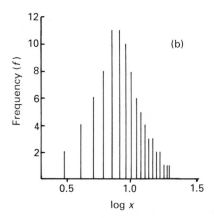

Fig. 10.2   (a) Untransformed observations; (b) logarithmically transformed observations.

discrete counts) but it has been *normalized* to an extent that the data may be used in parametric techniques without risk of serious error. Also, the dependence of the variance upon the mean has been removed.

Example 10.1

Five collections of moth larvae (caterpillars) are obtained by means of a beating tray from randomly selected locations in an oak wood. The numbers of larvae obtained are

28   5   17   31   77

Calculate the sum of squares of the transformed counts.

1. Transform the counts by replacing each observation x by $\log x$:

x	28	5	17	31	77
$\log x$	1.447	0.699	1.230	1.491	1.886

2. Use a 'scientific' calculator to obtain Σx^2 and $(\Sigma x)^2$ for the transformed values. These are 9.875 and 45.603, respectively. The sum of squares (SS) is given by:

$$SS = \Sigma x^2 - \frac{(\Sigma x)^2}{n} = 9.875 - \frac{45.603}{5} = 0.754$$

[*Hint*: a quicker way of obtaining a sum of squares with a calculator is to obtain the variance, s^2, and multiply it by the degrees of freedom, $n - 1$. In this example a calculator gives the standard deviation as 0.43439. The sum of squares is therefore, $0.43439^2 \times 4 = 0.754$.]

10.3   Logarithmic transformation – when there are zero counts

It is not uncommon to find observations of zero in a sample of count data. Such is the case with the orchid data in Example 8.3. Since the log of zero is meaningless, how do we apply the logarithmic transformation to these data? The answer is to add 1 to each observation before transforming. That is, replace x by $\log(x+1)$. Zero becomes 1; 1 becomes 2 and so on.

Example 10.2

Obtain the transformed mean and variance of the sample of orchid counts given in Example 8.3. The counts are: 0, 0, 4, 2, 58, 1, 0, 22, 5, 7, 17, 1. (In Example 8.3 we determined that the mean and variance were 9.75 and 281, respectively.)

1. Transform the counts by $\log(x+1)$:

x	0	0	4	2	58	1	0	22	5	7	17	1
$\log(x+1)$	0	0	0.699	0.477	1.771	0.301	0	1.362	0.778	0.903	1.255	0.301

2. Use a calculator to confirm that the mean of the transformed counts is 0.654 and the standard deviation is 0.5841. Therefore the variance is $0.5841^2 = 0.3412$. Notice how the transformation has reduced the value of the variance below that of the mean. This is good evidence that the transformation has been effective.

10.4   The square root transformation

The square root transformation is appropriate when the variance of a sample of count data is about equal to the mean, or when a Poisson distribution is suspected. In this transformation each observation x is replaced by its square root; that is, replace x by $\sqrt{x}$. If there are any zero counts, replace each observation by $\sqrt{x+0.5}$.

Example 10.3

A birdwatcher counts the number of Arctic skuas flying past a headland in 10 sampling units of 15 min. The observations are:

4   6   3   3   9   4   6   2   5   2

Use a calculator to find the variance of the square root transformed data.

1. Replace x by $\sqrt{x}$:

x	4	6	3	3	9	4	6	2	5	2
$\sqrt{x}$	2	2.45	1.73	1.73	3	2	2.45	1.4	2.24	1.4

2. The standard deviation s of the transformed data is 0.5069. The variance is therefore $0.5069^2 = 0.2569$.

Example 10.4

Use a calculator to find the standard deviation of the square root transformation of the following counts:

6   2   3   0   1   5

1. Since there is a zero observation, all counts are transformed by $\sqrt{x+0.5}$,

x	6	2	3	0	1	5
$\sqrt{x+0.5}$	2.550	1.581	1.871	0.707	1.225	2.345

2. The standard deviation, s, is 0.6917.

0.5   The arcsine transformation

The arcsine transformation is appropriate for observations which are *proportions*. We said in Section 10.1 that distributions of count data cannot be symmetrical because the left-hand tail of the distribution is truncated by the theoretical minimum of zero. In distributions which are proportions, *both* tails are truncated because all values must lie on a scale with absolute limits of 0 and 1. Errors which might arise are greatest if the observations are grouped at one end of the scale (say, close to 0.1 or 0.9) and least if they are near the middle (0.5).

The arcsine transformation requires two steps. First, obtain the square root of x. Second, find the angle whose sine equals this value. On most calculators it is obtained by pressing the $\sin^{-1}$ (inverse sine) key.

Thus, if $x = 0.25$

$$\sqrt{x} = 0.5$$
$$\sin^{-1} 0.5 = 30°.$$

Therefore, arcsine $0.25 = 30°$

Notice that the scale of the transformation is angular *degrees*. If observations are percentages, divide each by 100 before transforming.

Example 10.5

In an investigation into the effects of industrial pollution in the North Sea, seven trawls (sampling units) of fish are obtained. The proportion of flounders with skin lesions in each trawl unit is:

0.25   0.31   0.21   0.24   0.30   0.29   0.22

Use a calculator to obtain the variance of the arcsine transformed data.

1. Replace each x by arcsine x:

x	0.25	0.31	0.21	0.24	0.30	0.29	0.22
arcsine x	30.0	33.8	27.3	29.3	33.2	32.6	28.0

2. Using a calculator we find that the standard deviation s of the transformed observations is 2.6045 and therefore the variance s^2 is 6.7834.

10.6   Back-transforming transformed numbers

Usually we transform data to normalize them and to stabilize the variance so that they may be used in parametric techniques which require these conditions. Nearly always a test is performed upon the transformed data and inferences may be drawn from the outcome of the test. In certain techniques, however, the object is to estimate the number of items in some population. In such cases, a transformed number may be difficult to interpret – many biologists would be hard pressed to think of flounders in angular degree units! An example is given in Section 11.5. There, the number is restored to the original scale by the process of **back-transforming**. This is simply a reversal of transformation.

(a) *Log x transformation*:
   To back-transform 1.380, obtain the antilog.
   Antilog $1.380 = 24$.

(b) *Log (x + 1) transformation*:
   To back-transform 1.0414, take the antilog and subtract 1.
   (antilog $1.014) - 1 = 10$.

(c) *Square root x transformation*:
   To back-transform 3.464, obtain the square.
   $3.464^2 = 12$.

(d) *Square root* $(x+0.5)$ *transformation*:
To back-transform 3.937, obtain the square and subtract 0.5.
$3.937^2 - 0.5 = 15$.

(e) *Arcsine transformation*:
To back-transform 28.2, obtain the sine and square.
$(\sin 28.2)^2 = 0.2233$.

0.7 Is data transformation really necessary?

Users of statistics are sometimes accused of 'fiddling' or 'massaging' data if their observations do not at first support their predictions. Is the transformation of data by the methods we have outlined a dubious form of statistical massage? The answer is a firm *no*! First, data transformation is a well-established mathematical practice; the units of acidity (pH), noise (decibels) and seismic activity (Richter units) are three well-known examples of transformed scales. Second, in statistics the transformation of data is undertaken *a priori*; that is to say, it is planned in advance as part of the statistical technique. Transformation is not used as an afterthought (*a posteriori*) to change an initial 'unfavourable' result. If it is, then this *is* cheating! Transformation merely ensures that a particular statistical method can be validly applied.

Although the process of transforming data is not difficult, with large data sets it can be laborious and repetitive even with a calculator. The question therefore arises as to whether it is worth the effort. If no more is required than a simple statement of a sample mean and standard deviation or variance, transformation is not expected. But we reiterate, for valid use some statistical tests require that data are normal and with stable variance. If data are badly skewed, and therefore do not meet these requirements, then errors of unpredictable magnitude may arise if the data are not first transformed.

However, there *is* an alternative to data transformation. The non-parametric 'distribution-free' techniques we describe later in the text do not assume or require normally distributed data and can be safely used without transformation.

11 HOW GOOD ARE OUR ESTIMATES?

11.1 Sampling error

Usually we obtain a sample in order to derive a statistic (a mean, for example) from the observations that constitute the sample. The sample statistic gives an estimate of the corresponding population parameter; thus $\bar{x}$ estimates μ.

Intuitively, we expect that large samples provide more reliable estimates of parameters than small samples and, conversely, that small samples are less reliable than large ones. Several small samples drawn from the *same* population generally provide different values of the same statistic, yet they are all estimates of the same population parameter. The variation between these individual estimates is due to **sampling error**.

Sampling error arises because some samples have, by chance, more than a 'fair share' of larger units whilst others have more than a 'fair share' of smaller units. Sampling error is not a mistake or error due to an observer; rather it reflects the random scatter inherent in any sample. In a collection of small samples drawn from the same population, some values of a statistic underestimate the parameter and others overestimate it. The way in which sample statistics cluster around a population parameter is called the **distribution of the statistic** or, sometimes, the **sampling distribution**. Distributions of sample statistics conform to mathematical principles which allow us to state the confidence we may place in estimates of particular population parameters.

11.2 The distribution of a sample mean

The 100 observations of shoot lengths in Section 3.10 are presented again in Table 11.1 together with the means of each row. The grand mean of the whole sample has already been worked out as 74.00 mm. Scrutiny of the 10 row means shows that they vary, ranging from 72.7 mm to 74.7 mm, with not a single one equal in value to the grand mean.

We can group the row-means into a frequency distribution just as we grouped the observations themselves into a frequency distribution in Fig. 4.7.

Frequency class	Tallies	Frequency f
72.5 – 73.4	11	2
73.5 – 74.4	111111	6
74.5 – 75.4	11	2

Table 11.1 100 shoot-length measurements

				Shoot length (mm)						Mean $\bar{x}$
76	73	75	73	74	74	74	74	74	77	74.4
74	72	75	76	73	71	73	80	75	75	74.4
68	72	78	74	75	74	69	77	77	72	73.6
72	76	76	77	70	77	72	74	77	76	74.7
78	72	70	74	76	72	73	71	74	74	73.4
75	79	75	74	75	74	71	73	75	73	74.4
75	70	73	75	70	72	72	71	76	73	72.7
74	76	74	75	74	76	75	75	73	73	74.5
78	74	73	75	74	73	72	76	73	76	74.4
74	71	72	71	79	78	69	77	73	71	73.5

Although there are only 10 of these sub-set means, a clear, symmetrical pattern begins to emerge. Had we more sub-set means drawn from a larger sample, their distribution would resemble that shown in Fig. 9.1(a). Because the sample mean is a continuous variable, the distribution verges towards a smooth continuous curve, that is, the normal curve of distribution. This conclusion follows from a fundamental principle in statistics:

The Central Limit Theorem *states that the means of a large number of samples drawn randomly from the same population are normally distributed and the 'mean of the means' is the mean of the population.*

This normal distribution has its own standard deviation, that is, a **standard deviation of sample means**. The standard deviation of a sample means is given its own name: the **standard error of the mean**.

It is important to note that the Central Limit Theorem makes no assumptions about the underlying distribution of the population from which samples are drawn. That distribution does *not* have to be normal; it may be symmetrical, skewed or bimodal; observations may be continuous or discrete. Whichever is the case, the means of a large number of samples are approximately normally distributed around the population mean, μ.

The properties of the normal curve described in Chapter 9 hold true for a normal distribution of sample means. Thus, about 68% of a large number of sample means fall within ± 1 standard error (S.E.) of the population mean μ. The converse of this is also true: we are similarly confident (68%) that a *population* mean μ falls within ± 1 S.E. estimated from a *sample* mean $\bar{x}$.

In practice, we do not estimate a standard error by looking at the spread of sub-sets of sample means. Instead, it may be calculated from the observations of a sample by:

$$S.E. = \frac{\text{sample standard deviation}}{\sqrt{\text{number of observations}}} = \frac{s}{\sqrt{n}}$$

Example 11.1

Calculate the standard error of the mean of the 100 shoot length measurements for which we have already worked out the mean (74.00 mm) and standard deviation (2.34 mm) in Section 6.5.

$$\text{S.E.} = \frac{s}{\sqrt{n}} = \frac{2.34}{\sqrt{100}} = \frac{2.34}{10} = 0.234$$

In plain language this means: the mean shoot length of the sample is 74.00 mm and the standard error of the mean is ± 0.234 mm. We are therefore 68% confident that the mean of the *population* lies between $74.00 + 0.234$ ($= 74.234$ mm) and $74.00 - 0.234$ ($= 73.766$ mm). This is, of course, more informative than the standard deviation (s) because it indicates how close the sample mean is likely to be to the population mean which is what we seek to estimate.

11.3   The confidence interval of the mean of a large sample

The standard error of the mean gives us an indication of how good an estimate a sample mean, $\bar{x}$, is of a population mean, μ. Thus, we are 68% confident that μ lies within ± 1 S.E. of $\bar{x}$. However, 68% is a rather low level of confidence; we usually want to be surer that a population mean lies between indicated limits. To meet this need, 95% or 99% limits are generally used. These can be obtained by multiplying the standard error by the appropriate z-score as follows:

- we are 95% confident that a population falls within ± 1.96 S.E. of a sample mean
- we are 99% confident that a population mean falls within ± 2.58 S.E. of a sample mean

1.96 and 2.58 are the same z-scores that are used in describing the properties of the normal curve (Section 9.2) and are valid only in samples containing 30 or more observations. The intervals $\bar{x} \pm 1.96$ S.E. and $\bar{x} \pm 2.58$ S.E. are called **95%** and **99% confidence intervals**, respectively. The adjustments made in the case of small samples are described in Section 11.4.

Example 11.2

Estimate the 95% confidence interval of the mean of the 100 shoot length measurements given in Section 11.2.

We have determined in Example 11.1 that the mean of the sample is 74.00 mm and the standard error is ± 0.234 mm. The 95% confidence interval is therefore $74.00 \pm (1.96 \times 0.234) = 74.00 \pm 0.459$ mm. This means we are 95% confident that the population mean lies between 74.459 mm and 73.541 mm.

Notice that because n, the number of observations, is the denominator of the equation for estimating the standard error (and hence confidence interval) the value of the standard error (and the breadth of the confidence interval) gets smaller as n gets larger. This is a mathematical expression of the statistical axiom that the larger the sample size, the greater is the reliability of an estimate of a population parameter.

1.4 The confidence interval of the mean of a small sample

In calculating the standard error of a mean (and hence a confidence interval) the standard deviation is used. Strictly, this should be the population standard deviation, σ. In large samples we are confident that the sample standard deviation, s, is a reliable estimate of σ; we are less certain however in the case of small samples. We therefore need to apply a correction factor to compensate for the uncertainty in small samples. That factor should become larger as the sample becomes smaller. A suitable correction factor is t, as described in Section 9.5.

Example 11.3

Calculate the mean, with 95% confidence limits of a sample of observations of small mammal masses (g):

19.4   21.4   22.3   22.1   20.1   23.8   24.6   19.9   21.5   19.1

Using a scientific calculator it is determined that $\bar{x}=21.42$ and $s=1.84$. The standard error is:

$$\text{S.E.} = \frac{s}{\sqrt{n}} = \frac{1.84}{\sqrt{10}} = 0.582$$

In a small sample we do not use the z-score of 1.96 to obtain the confidence interval but the value of t found in tables against the appropriate number of degrees of freedom $(n-1)$. Thus:

95% confidence interval $= \bar{x} \pm (t \times \text{S.E.})$

In Appendix 2 we find that for $(10-1)$ degrees of freedom at $P=0.05$ (95%), $t=2.262$. The 95% confidence interval is therefore:

$21.42 \pm (2.262 \times 0.582)$, i.e. $21.42 \pm 1.32\,\text{g}$

Thus, we are 95% confident that the mean mass of the population from which the sample is drawn lies between 22.74 g (upper limit) and 20.1 g (lower limit). If we had used the z-score of 1.96 instead of the t-score of 2.262, the limits would have been 22.56 g and 20.28 g – an important reduction of range of about a third of a gram.

11.5   The confidence interval of the mean of a sample of count data

Because the Central Limit Theorem states that a set of sample means drawn from *any* single population is normally distributed about the population mean, μ, the calculation of a 95% confidence interval of the mean of a large sample of observations of counts is exactly the same as described in Section 11.3. However, because samples of count data are often drawn from populations which are greatly skewed, the application of the factor *t* may not be a sufficient correction in the case of a small sample. As a precaution, the confidence interval about the mean of a small sample of count data should be calculated upon transformed observations. We described the appropriate transformations for random, contagious and proportional data in Chapter 10. Because contagious dispersion is most frequently encountered, we choose this as our example to illustrate the technique for estimating a confidence interval.

Example 11.4

Recall Example 8.3 in which counts of green-winged orchids in a sample of twelve 1×1 m quadrats placed randomly in a hay-meadow were presented:

0   0   4   2   58   1   0   22   5   7   17   1

Because the variance of the sample is very much larger than the mean, a logarithmic transformation is applicable. Note that there are some zero observations; we therefore use the transformation $\log(x+1)$.

Transforming the observations x to $\log(x+1)$ we obtain:

0   0   0.699   0.477   1.771   0.301   0   1.362   0.778   0.903   1.255   0.301

The mean of the transformed counts is 0.654 and the standard deviation is 0.584.

The standard error is $\dfrac{s}{\sqrt{n}} = \dfrac{0.584}{\sqrt{12}} = 0.169$

The 95% confidence interval of the transformed counts is therefore:

$0.654 \pm (t \times 0.169)$ where $t(0.05)$ for 11 df is 2.201.

The interval is:

$0.654 \pm (2.201 \times 0.169)$

$= 0.654 \pm 0.372$

Back-transforming by taking antilogs and subtracting 1, this becomes

$3.508 \overset{\times}{\underset{\div}{}} 1.355$

[*Note*: a number which is added and subtracted as a logarithm becomes multiplied and divided when antilogged.]

The number 3.508 is the derived mean, $\bar{y}$ and is considerably smaller than the simple arithmetic mean of 9.75 worked out in Example 8.3. It is known as the **geometric mean** and, if used in population estimates, results in corresponding underestimation. There are mathematical techniques available to correct for this; however, a satisfactory practical approximation is to use the **arithmetic mean** as the estimate of μ but the *limits derived by transformation* for setting the confidence interval. Thus, in the present example, we are 95% confident that if we were able to count the orchids in every possible quadrat in the meadow (i.e. the statistical population) the mean number of orchids per quadrat would fall between:

$9.75 \times 1.355 = 13.21$ (upper limit)

and

$9.75 \div 1.355 = 7.196$ (lower limit)

This interval has practical application if we wish, for example, to compare several meadows. Sometimes, however, biologists obtain data from a sample in order to estimate a 'biological' population – that is, to answer the question, 'How many orchids are there in the meadow?'. If we say that the area of the meadow is 500 m², the crude estimate of the total number of orchids is the sample mean (untransformed) × N (the total number of all possible sampling units). That is, $9.75 \times 500 = 4875$ orchids. For practical purposes, the confidence interval of this estimate is based upon the confidence interval obtained from the *transformed* counts, that is $13.21 \times 500 = 6605$ (upper limit) and $7.196 \times 500 = 3598$ (lower limit). Notice that the confidence interval is not spaced symmetrically above and below the mean. This is, of course, consistent with a very skewed distribution (Fig. 11.1).

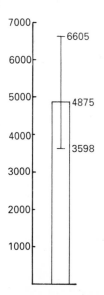

Fig. 11.1   Confidence interval of the estimated number of orchids in a field.

In similar fashion, a sample of counts obtained from a randomly dispersed population (Poisson distribution) is treated by means of a square root transformation. Regularly dispersed data are harder to deal with accurately but providing that p is between about 0.4 and 0.6 and k is larger than about 10, transformation is not needed. If the estimated values of p and k are not within these limits, a 'conservative' transformation is the square root. That is to say, treat the values as if they are randomly dispersed. Observations which are proportions are transformed by means of the arcsine transformation.

11.6   The difference between the means of two large samples

Field biologists often compare the means of some variable in two samples. To know how much heavier, larger, faster, warmer or more acidic one sample is than another allows us to make inferences about differences between the populations from which the samples are drawn.

Example 11.5

Conservationists studying the skink *Scelotes bojerii* (a sort of lizard) on small islands off the coast of Mauritius observe that samples captured on Gunner's Quoin are on average smaller than those on Round Island. They suggest that rats, which have been introduced to Gunner's Quoin, are selectively predating the larger animals, thus reducing the average size. In order to quantify the difference in size, a large sample of skinks from each island is obtained and measured (snout-to-vent length is a suitable index of size). The results are tabulated below.

Gunner's Quoin: $\bar{x} = 35.33$ mm; $s = 7.124$
Round Island: $\bar{x} = 56.50$ mm; $s = 7.714$ $\Big\} n = 30$ in each case

Employing the technique described in Section 11.3, we estimate the 95% confidence interval (C.I.) for each mean, using the t-score of 2.045 for a sample size of 30 $(= 29$ df$)$:

Gunner's Quoin: 95% C.I. $= 35.33 \pm \left(2.045 \times \dfrac{7.124}{\sqrt{30}} \right) = 35.33 \pm 2.66$ mm

Round Island: 95% C.I. $= 56.50 \pm \left(2.045 \times \dfrac{7.714}{\sqrt{30}} \right) = 56.50 \pm 2.88$ mm

These results are expressed graphically in Fig. 11.2. The difference between the two sample means is $(56.50 - 35.33) = 21.17$ mm. We record: 'the mean of the sample obtained from Round Island is 21.17 mm larger than that obtained from Gunner's Quoin. But what are we to infer about the difference in size between the two populations from which the samples are drawn?

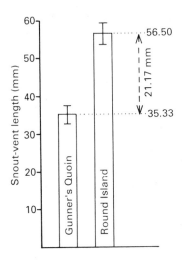

Fig. 11.2 The difference between two sample means.

The problem of course is that we only have estimates of the two population means. We are 95% confident only that they fall within the respective intervals indicated in Fig. 11.2. It follows that the *difference* between the two sample means is also only an estimate of the difference between the two population means and that we need a method of attaching confidence limits to this estimate. If we were able to obtain measurements from all skinks on both islands we would then know the value of each population mean (μ_1 and μ_2). The real difference between the population means is ($\mu_1 - \mu_2$), the **population mean difference**. The difference between any pair of *sample* means ($\bar{x}_1 - \bar{x}_2$) is an estimate of ($\mu_1 - \mu_2$). If we could obtain several pairs of samples, we could accordingly obtain several values of ($\bar{x}_1 - \bar{x}_2$). Each would probably be different in value due to sampling error but each would nevertheless be an independent estimate of the absolute parameter ($\mu_1 - \mu_2$). Statistical theory predicts that a set of mean differences drawn from two populations clusters around ($\mu_1 - \mu_2$) in a manner which resembles a **normal distribution**. This normal distribution of mean differences has its own mean and standard deviation parameters: the mean is equal to the population mean difference, ($\mu_1 - \mu_2$), and the standard deviation is called the **standard error of the difference**.

By similar arguments that were used in the case of the standard error of the mean (Section 11.2) we are about 68% confident that a *population mean difference* falls within ± 1 S.E. of a *sample mean difference*. We now need an expression for estimating the standard error of the difference between two sample means. It is the square root of the sum of the squared individual sample standard errors:

$$S.E._{\cdot diff} = \sqrt{S.E._{\cdot 1}^2 + S.E._{\cdot 2}^2}$$

Thus,

$$\text{S.E.}_{\text{diff}} = \sqrt{\frac{(s_1^2 + s_2^2)}{(n_1 + n_2)}}$$

For the skink data (where $n = 30$ in each case):

$$\text{S.E.}_{\text{diff}} = \sqrt{\frac{7.124^2}{30} + \frac{7.714^2}{30}} = 1.917$$

Because the samples are regarded as large, we multiply the S.E.$_{\text{diff}}$ by the z-score of 1.96 to convert to an approximate 95% confidence interval. We are therefore 95% confident that the true population mean difference $(\mu_1 - \mu_2)$ lies between $(\bar{x}_1 - \bar{x}_2) \pm 1.96 \text{ S.E.}_{\text{diff}}$:

$$
\begin{aligned}
\text{True population mean difference} &= 21.17 \pm (1.96 \times 1.917) \\
&= 21.17 \pm 3.76 \text{ mm} \\
&= 24.93 \text{ mm (upper limit) and } 17.41 \text{ mm} \\
& \text{(lower limit)}
\end{aligned}
$$

This result is clearly a good deal more informative than the unprocessed sample mean difference of 21.17 mm.

We should note two points here. First, if the larger sample mean is nominated as $\bar{x}_1$ and the smaller as $\bar{x}_2$ then the sample mean difference is conveniently positive. Second, provided that *both* of the samples are quite large (usually considerably more than 30 observations) it does not matter if they are unequal. Remember, we are able to use the z-score of 1.96 because it is the normal distribution of *sample mean differences* about a *population mean difference* that applies – not the distribution of observations within the populations themselves. In critical work, the corresponding t-score for $(n_1 + n_2 - 2)$ degrees of freedom may be used in place of the z-score of 1.96. In this example, t_{58} for $P = 0.05 \simeq 2.000$ but its use extends the confidence interval by less than 0.1 mm.

11.7   The difference between the means of two small samples

The rationale for establishing a confidence interval about the difference between the means of two small samples is the same as for that of large samples. The estimation of the standard error of the difference is a little more difficult, however. The expression for the standard error of the difference is:

$$\text{S.E.}_{\text{diff}} = \sqrt{\left[\frac{(n_1 - 1)s_1^2 + (n_2 - 1)s_2^2}{(n_1 + n_2 - 2)}\right]\left[\frac{(n_1 + n_2)}{(n_1 n_2)}\right]}$$

Although this is a rather cumbersome equation, the terms within it are familiar: they are simply the sample size and standard deviation of each of the two samples being considered.

Example 11.6

Estimate the difference between the means of the two samples of skink measurements for which the sample statistics are given in Example 11.5, if the sample sizes are now 12 and 10 for Gunner's Quoin and Round Island, respectively.

The difference between the sample means is 21.17 mm, as before. Using the equation above (where n_1, s_1 and n_2, s_2 refer to Gunner's Quoin and Round Island, respectively):

$$
\begin{aligned}
\text{S.E.}_{\text{diff}} &= \sqrt{\left[\frac{(11 \times 7.124^2) + (9 \times 7.714^2)}{20}\right]\left[\frac{(12 + 10)}{120}\right]} \\
&= \sqrt{\frac{(558.27 + 535.55)}{20} \times 0.1833} \\
&= \sqrt{10.025} \\
&= 3.166
\end{aligned}
$$

Because the samples are small, the use of the z-score of 1.96 is not appropriate; t at $(n_1 + n_2 - 2)$ degrees of freedom is used in its place. For $P = 0.05$, t_{20} is 2.086. The confidence interval is therefore:

$$
\begin{aligned}
\text{C.I.} &= 21.17 \pm (2.086 \times 3.166) \\
&= 21.17 \pm 6.60 \text{ mm} \\
&= 27.77 \text{ mm (upper limit) and } 14.57 \text{ mm (lower limit)}
\end{aligned}
$$

This of course represents a considerable extension of the confidence interval estimated for the larger samples of Example 11.5.

1.8 Estimating a proportion

Field biologists often express the frequency of occurrence of an item in a sample as a proportion of the total. For example, a sample of mosquito larvae collected in a pond net contains 80 larvae of which 60 are *Aedes detritus*. The proportion of that species in the sample is $60/80 = 0.75$ (75%). Provided that the larvae are dispersed independently, this proportion is an estimate of the actual proportion of *A. detritus* in the population. Subsequent samples are all, in turn, independent estimates of the same population proportion but, due to sampling error, are likely to be different in value from each other. In the same way that a set of sample means is distributed around a population mean, so too is a set of sample proportions distributed around a population proportion. The standard deviation of the distribution is similarly called the **standard error**. For practical purposes, a satisfactory estimate of the standard error from a sample proportion is given by:

$$S.E. = \sqrt{\frac{p(1-p)}{(n-1)}}$$

where p is the proportion of the nominated item and n is the number of all items in the sample.

Example 11.7

Estimate the standard error and 95% confidence interval of the proportion of *Aedes detritus* when 60 out of 80 larvae in a sample are of this species.

$$S.E. = \sqrt{\frac{0.75(1-0.75)}{(80-1)}} = 0.049$$

The limits of the confidence interval are obtained by multiplying the standard error by 1.96. The interval is:

$0.75 \pm (1.96 \times 0.049)$

$= 0.75 \pm 0.096$

That is, 0.846 (84.6%) upper limit and 0.654 (65.4%) lower limit. We are therefore 95% confident that the true proportion of *A. detritus* lies between 0.654 and 0.846 (65.4% and 84.6%).

Two points should be noted here. First, dramatic increases in sample size are required to reduce the confidence interval by an appreciable extent. Thus, increasing the sample size from 80 to 200 in Example 11.7 reduces the standard error from 0.049 to 0.03, a reduction of only 0.019. Second, the formula for calculating the standard error becomes unreliable when p is less than 0.1 or greater than 0.9.

11.9   Estimating a Lincoln Index

The **Lincoln Index** is an estimate of the number of individuals in a population based on the proportion of individuals marked in a first sample which are present as recaptures in a second sample.

Example 11.8

A sample (n) of 200 grasshoppers captured by sweep-netting a meadow are marked with a spot of paint and released back into the meadow. After allowing an interval for the insects to disperse, a second sample (N) of 450 grasshoppers is obtained of which 15 (r) are recaptures from the first sample. What is the estimate of the number of grasshoppers in the meadow?

From the second sample, we deduce that for every 450 individuals in the population, 15 are marked. Therefore every single marked individual represents $450/15 = 30$ individuals in the population. Since we know there are 200 marked individuals, the population estimate is $30 \times 200 = 6000$ grasshoppers. The general relationship is given by:

$$\text{Lincoln Index} = \frac{N \times n}{r}$$

This is, of course, an estimate and the number of marked individuals in the second sample is subject to sampling error. However, we would expect a set of replicate 'second samples' to generate a set of estimates which are distributed around the true population size. This distribution has a standard deviation called a **standard error**. It is estimated from:

$$\text{S.E.} = \sqrt{\frac{n^2 \times N(N-r)}{r^3}}$$

Using the above data this becomes:

$$\text{S.E.} = \sqrt{\frac{200^2 \times 450(450-15)}{15^3}}$$

$$= \sqrt{\frac{40\,000 \times 195\,750}{3375}}$$

$$= 1523$$

Multiplying the standard error by the z-score of 1.96, we estimate the 95% confidence interval about the population estimate, thus:

$$\text{Number of grasshoppers} = 6000 \pm (1.96 \times 1523)$$
$$= 6000 \pm 2985$$
$$= 8985 \text{ (upper limit) and } 3015 \text{ (lower limit)}$$

In other words, we are 95% confident that the meadow contains between 3015 and 8985 grasshoppers. We emphasize that we have shown only the arithmetic and statistical treatment of the sample data. Associated with these are a number of *experimental* assumptions underlying the technique. These include: negligible mortality during the experimental period; complete integration of the released marked individuals among the unmarked; negligible immigration into and emigration out of the population. For a full appraisal of these assumptions, and for a review of other mark-recapture methods, readers are referred to Southwood (1978).

.10 Estimating a diversity index

Measures of species diversity have application in conservation assessment, it being argued that sites with high diversity are more valuable than those with low diversity. The number of species present ('species richness') is a simple index of diversity but there are a number of other indices which take into account how evenly the total number of individuals in a sample is apportioned between each species ('equitability'). One of the most widely used indices is that known as the **Shannon–Weaver Index** which is defined as:

$$H = -\sum_{}^{s} p_i \ln p_i$$

The term p_i is the proportion of a particular species in a sample which is multiplied by the natural logarithm of itself. H is derived by summing the product for all species in the sample. The minus sign is to make the final value of H positive.

Example 11.9

A sample of moths captured in a moth trap contained six species with the following numbers of each species:

55   30   15   10   5   3

Calculate the Shannon–Weaver diversity index for this sample.

The total number of moths in the sample is 118.
The index is therefore calculated by:

$$-\left[\left(\frac{55}{118}\right) \times \ln\left(\frac{55}{118}\right)\right] + \left[\left(\frac{30}{118}\right) \times \ln\left(\frac{30}{118}\right)\right] \cdots + \left[\left(\frac{3}{118}\right) + \ln\left(\frac{3}{118}\right)\right]$$

$$H = -(-1.403) = 1.403$$

This value is of course an estimate and we would expect that a set of such sample estimates is distributed about the true population value. Unfortunately, there is no simple method of estimating the standard error from sample data. Readers with a more advanced knowledge of mathematics may consult Pielou (1975) but, for practical purposes, a diversity index can be regarded as a number on an *ordinal* scale. Thus, an index of 4.0 does *not* mean 'twice as diverse' as an index of 2.0. There are statistical techniques available for dealing with ordinal data. For example, a set of diversity indices obtained from several estimates at one site may be compared with a set of a similar number of estimates from another site by means of the Mann–Whitney U-test (Section 16.3). Notice here the word 'sample' has to be applied thoughtfully. A diversity index is estimated from a sample of organisms. However, a set of such indices itself constitutes a sample which may be compared with another sample of indices.

Two cautionary points need to be made. First, it is sensible to use diversity indices only to compare 'like with like'. The diversity index of moths in a sample may reasonably be compared with an index of moths in a similar sample from another site, but not with an index of birds. Second, comparisons between diversity indices are valid only if there are approximately equal numbers of organisms in the samples from which the indices are calculated.

.11 The distribution of a variance – chi-square distribution

The distribution of a sample variance, s^2, is unlike that of a sample mean in that it is distributed asymmetrically about the population parameter σ^2. The left-hand side of the distribution is truncated at the minimum value of zero when all observations in a sample by chance have identical values but the right-hand side may in theory extend to infinity. This introduces a positive skew to the distribution. Like the distribution of t, the shape of the distribution of a variance depends on the sample size or, more precisely, the degrees of freedom, $(n-1)$. The larger the size of the replicate samples, the more symmetrical becomes the distribution and for very large samples ($n > 100$) the distribution converges towards normality. Figure 11.3(a) shows the approximate shapes of the distributions of sample variances for 1, 3 and 10 degrees of freedom.

These distributions may be re-scaled by dividing the vertical axis by the total number of observations in the distribution in which case they are effectively converted to probability rather than frequency distributions – see Section 7.3. If we then standardize the horizontal axis by multiplying the variance by the degrees of freedom, thus converting it to a sum of squares (see Section 6.4) and then dividing by the population variance σ^2 we generate a new family of probability distributions (or, strictly, *densities*) known as χ^2 (chi-square) distributions. There is a separate distribution for each possible number of degrees of freedom. Figure 11.3(b) shows a χ^2 probability density for 9 degrees of freedom.

Like all probability densities, the area under the curve represents 100% of the probability. We are interested in the point on the horizontal axis which excludes 5% of the probability. Nearly always we use χ^2 in *one-tailed* tests (the reason for this is explained in Section 13.1). We therefore look for the point which excludes 5% in the right hand tail (Fig. 11.3b). The required value for a particular number of degrees of freedom is found in tables.

Example 11.10

Find the critical value of χ^2 for $P = 0.05$ at 9 degrees of freedom.

Enter Appendix 3 at 9 df in the column headed 0.05. We find the value 16.92. A value which is larger than this (at 9 df) is regarded as 'unlikely', or *statistically significant*.

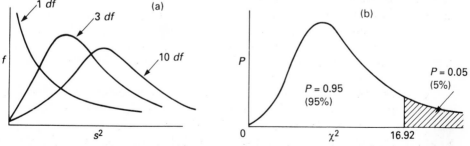

Fig. 11.3 (a) Distributions of sample variances for 1, 3 and 10df; (b) distribution of χ^2 at 9df showing upper 5%.

12 THE BASIS OF STATISTICAL TESTING

12.1 Introduction

In Section 7.8 we said that statisticians set arbitrary critical thresholds of probability. When an event occurs whose probability is estimated to be below a critical threshold, the outcome is said to be *statistically significant*. The critical values are: $P < 0.05$ (*significant*); $P < 0.01$ (*highly significant*); and $P < 0.001$ (*very highly significant*).

Example 9.1 shows that we can use the properties of the normal curve to estimate probabilities. Thus, an observation obtained randomly from a normally distributed population and having a value exceeding the population mean (μ) plus or minus 1.96 standard deviations occurs in fewer than 1 trial in 20. That is to say, the probability of such an outcome is $P < 0.05$. If a single random observation does exceed this critical value it is regarded as being statistically significant. The procedure for deciding if the outcome of a particular event is significant is called a **statistical test**. We now need to explain in more detail the basis of statistical testing.

12.2 The experimental hypothesis

The formulation and testing of hypotheses is the basis of experimental science. A hypothesis is a proposed explanation for a state of affairs. A hypothesis is tested by experimentation, the outcome of which may provide evidence for the acceptance or rejection of the hypothesis. As an example of an experimental hypothesis we refer to the reptiles inhabiting two islands (Round Island and Gunner's Quoin) described in Section 11.6.

The state of affairs: Skinks are, on average, larger on Round Island than on Gunner's Quoin.

Experimental hypothesis: Rats, which are present only on Gunner's Quoin, selectively predate the larger skinks, thus reducing the average size.

Possible experiment: Analyse the diets of rats on Gunner's Quoin to determine whether they contain a high proportion of larger skinks.

If it is found that rats *are* selectively predating larger animals the hypothesis is accepted for the time being. The outcome of the experiment does not *prove* that rats are responsible for the size discrepancy; it simply fails to provide evidence for doubting it. On the other hand, if the experiment shows that rats predate large and small skinks equally, the hypothesis has to be rejected, and an alternative proposed.

12.3 The statistical hypothesis

In our reptilian example above the description of the *state of affairs* is based on the observation that the mean size of a sample of skinks from Round Island is greater than that of a sample from Gunner's Quoin. Could there be any reason to doubt the validity of this difference? To examine this suggestion we nominate the mean size of the population from which the Round Island sample is drawn as μ_1, and that from Gunner's Quoin as μ_2.

Let us assume for the moment that there is *no* difference between the means of the two populations, and $\mu_1 = \mu_2$. (This is actually called a Null Hypothesis, explained below.)

If the population means are identical, we can superimpose the size frequency distribution of one population over that of the other, as we have done in Fig. 12.1. Furthermore, if we assume that there are approximately equal numbers of skinks on each island, then the two distributions will have nearly the same height.

Now imagine that a random sample is drawn from each population. Although each unit (a skink) is drawn randomly it is possible, by chance, that one sample has more than a 'fair share' of larger units. This is due to *sampling error*, described in Section 11.1. If, also by chance, the other sample has more than a 'fair share' of smaller units, then there could be a substantial difference

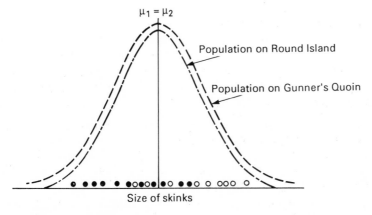

Fig. 12.1 Size frequency distributions of skinks on two islands. •, random observations from Round Island; ○, random observations from Gunner's Quoin.

between the means of the samples, *even though they are drawn from populations with identical means.*

The likelihood of such a spurious event arising diminishes rapidly as the number of units in each sample increases. A statistical test is required to determine the probability that two samples of given size and with a particular mean difference could have been drawn from two populations with identical means. If the calculated probability is below the critical threshold, we accept the evidence and conclude that the means of the two samples are statistically significantly different and have been drawn from populations with different means. The basis of a statistical test is a hypothesis that presumes there is *no* difference. This is called a **Null Hypothesis** and is symbolized by H_0. An alternative to a Null Hypothesis is symbolized by H_1. Usually, an alternative hypothesis is simply that there *is* a difference; for example $\mu_1 \neq \mu_2$.

By analogy with the experimental hypothesis of Section 12.2 we write:

State of affairs: The mean length of a sample of skinks from Round Island is greater than that from Gunner's Quoin.

Null Hypothesis H_0:   Although the sample means are different, we assume for the moment that the two samples have been drawn from two populations which have identical means, i.e. $\mu_1 = \mu_2$. If H_0 is false then $\mu_1 \neq \mu_2$.

Experiment   A statistical test which computes the probability that two
(statistical test):   samples of their particular sample size and mean difference could have been drawn from two populations with identical means.

If the outcome of the test suggests that the probability of the two samples being drawn from two populations with identical means is too low to be acceptable, we reject H_0 and accept our alternative hypothesis, H_1. If this is the case, we may confidently proceed with our fieldwork to establish the cause of the difference, assured that there is little risk that our evidence is based upon spurious sampling error.

12.4 Test statistics

The objective of each of the statistical tests that we shall describe is to produce a single number called a **test statistic**. The important feature of a test statistic is that its probability distribution is known, having been worked out by statisticians; z, whose probability distribution we outlined in Section 9.3 is such a statistic. For example, we note in Section 9.3 that the probability (P) of an observation falling outside a z-value of ± 1.96 is less than 0.05; such an observation is therefore considered *significant*. In small samples the same consideration applies to the appropriate value of t.

In each case, a statistical test produces a value for its test statistic, and the task is to determine whether that value exceeds some probability threshold which suggests rejection of a Null Hypothesis. Fortunately, published tables of the threshold values of each test statistic are readily available and, as we shall show, are easy to use. If a test is performed on *transformed* data, the test statistic is *not* transformed back to the original scale.

12.5   One-tailed tests and two-tailed tests

In Section 9.3 we distinguished between one-tailed tests and two-tailed tests. Now we may relate these to the idea of hypothesis testing.

In our example in Section 12.3 we establish a Null Hypothesis that two samples are drawn from two populations with identical means. We make no prediction, if H_0 is false, as to which mean is larger than the other – only that they are different:

$$H_0 : \mu_1 = \mu_2 \qquad H_1 : \mu_1 \neq \mu_2$$

If the outcome of the test is such that we are obliged to reject H_0 and accept H_1 we are then entitled to conclude that the sample with the larger mean has been drawn from a population with a larger mean. This conclusion is only reached *after* the test, that is to say, *a posteriori*. A test like this which makes no prediction as to which mean is the larger of the two, should they prove to be different, is called a **two-tailed** test.

Sometimes it is possible to predict in advance that if two samples are not drawn from populations with the same mean, then a *nominated* sample has been drawn from a population with a larger mean or, alternatively, has been drawn from a population with a smaller mean. The Null Hypothesis is the same, but the alternative hypothesis is different. If the population predicted to have the larger mean is nominated μ_1 then:

$$H_0 : \mu_1 = \mu_2 \qquad H_1 : \mu_1 > \mu_2$$

One-tailed tests sometimes arise in taxonomy. For example, if a sample does not represent population A, then it must represent population B, which is known to be smaller.

Because a one-tailed test is less stringent than a two-tailed test, considerable caution should be exercised before using it. Moreover, 'significant' results can be obtained with smaller sample sizes. Observers who persuade themselves that a test is one-tailed in order to obtain a 'result' with a small sample are *definitely cheating*!

In summary, a one-tailed test should only be used when there is an *a priori* reason to predict a directional influence in the data; moreover, the decision to use a one-tailed test *must* be made before the data are analysed. If there is doubt – use a two-tailed test; an outcome which is statistically significant in a two-tailed test is also significant in a one-tailed test.

There is no difference between the execution of a one-tailed test and a two-tailed test; there is simply a lower threshold of significance in a one-tailed test. Two statistical tests which are always one-tailed are the chi-square test, and the F-test in analysis of variance. We explain why in due course.

12.6   Hypothesis testing and the normal curve

We may now relate the principles outlined in this chapter to the statistical tests undertaken in Chapter 9. Turn again to Example 9.1. There we ask if it is likely that a single observation of 4.3 mm is drawn randomly from a normal population for which $\mu = 3.8$ and $\sigma = 0.15$. Our Null Hypothesis H_0 is that the observation *is* drawn from such a population (H_1, that it is not). The test involves the computation of a test statistic z. The computed value of z (3.33) exceeds 1.96 which is the value corresponding to the critical probability threshold of $P = 0.05$. We therefore reject H_0, accept H_1 and conclude that the observation probably is not drawn from the population.

Look again at Example 9.2. In this case H_0 is that an observation of 6.1 is drawn from a normal population for which $\mu = 5.6$ and $\sigma = 0.26$. Because this is a one-tailed test, H_1 is that the observation is drawn from a population of which μ is larger than 5.6. Although the execution of the test is identical, we use the less stringent value of $z = 1.65$ to reject H_0. The computed value is 1.9, large enough to reject H_0 in the one-tailed test but not large enough to reject H_0 had the test been two-tailed.

Figure 12.2 illustrates how the normal curve is used in hypothesis testing.

12.7   Type 1 and type 2 errors

When the threshold for rejection of H_0 is set at $P = 0.05$, an investigator is said to *accept the 0.05 (or 5%) level of significance*. This means that in tests where the computed value of the test statistic is equal to, or barely exceeds, the critical value, the decision to reject H_0 is probably correct 19 times out of 20 or 95 times out of 100. It follows that 5 times out of 100 there is a risk of rejecting H_0 when it is *true*. When H_0 is rejected and it is actually true, we refer to a **type 1 error** having been committed. How can the risk of a type 1 error be reduced? Simply by setting the level of acceptance at a more rigorous standard, for example, at the 1 in 100 times level of significance ($P = 0.01$).

It will be appreciated however that the analyst faces a 'swings and roundabouts' situation. The opposite of the case just outlined is referred to as a **type 2 error**, that is, *not* rejecting the Null Hypothesis when, in fact, it should be rejected.

Thus, as the probability of making a type 1 error is reduced, the probability of making a type 2 error is increased.

In most statistical analyses the aim is usually to limit the probability of committing type 1 errors, thus erring on the side of caution. In practice, the calculated value of a test statistic often exceeds the tabulated critical value at $P = 0.05$ in which case we reject H_0 at $P < 0.05$ and the risk of error is accordingly reduced.

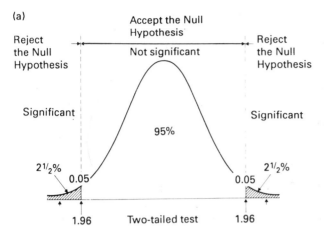

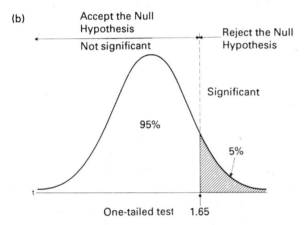

Fig. 12.2 Hypothesis testing and the normal curve: (a) two-tailed test; (b) one-tailed test.

2.8 Parametric and non-parametric statistics: some further observations

In Section 2.8 brief reference is made to the distinctions between parametric and non-parametric statistics. It is noted that the conditions relating to the use of parametric tests are more rigorous than those which apply to non-parametric tests. These distinctions are now elaborated upon in Table 12.1.

Biologists may wish to test nominal and ordinal level data in which case non-parametric tests are often very suitable. Non-parametric tests usually employ simple formulae and avoid the rather tedious computations of variances, sums of squares, etc. that are often required in parametric tests. Furthermore, they

Table 12.1 Distinctions between non-parametric and parametric tests

Non-parametric tests	Parametric tests
May be used with actual observations, or with observations converted to ranks	Are used only with actual observations
May be used with observations on nominal, ordinal and interval scales	Generally restricted to observations on interval scales
Compare medians	Compare means and variances
Do not require data to be normally distributed or to have homogeneous variance; i.e. they are 'distribution free'	Require data to be normally distributed and to have homogeneous variance
Are suitable for data which are counts	Counts must usually be transformed
Are suitable for derived data, e.g. proportions, indices	Derived data may first have to be transformed

require no data transformation. On the other hand, because non-parametric tests may be used with smaller samples, do not use *all* the data (that is **ranks** rather than the actual observations are used), and are more 'permissive', they may be less powerful than a corresponding parametric test.

12.9 The power of a test

Statisticians sometimes refer to the **power of a test**. It is a measure of the likelihood of a test reaching the correct conclusion, i.e. accepting H_0 when it should be accepted, and rejecting it when it should be rejected. Non-parametric tests are generally regarded as being less powerful than parametric equivalents, because they are less rigorous in their conditions of use. It must be emphasized, however, that parametric tests are more powerful than non-parametric only when the assumptions governing their use hold true. If an assumption does not hold true (e.g. the data do not fit a normal distribution), it is not just a question of a reduction of power; the whole validity of the test is destroyed and there may be the risk of considerable error.

There is one safe rule:

If there is doubt as to whether a particular set of data satisfies the assumptions made in the use of a parametric test, then a non-parametric alternative should be used.

3 ANALYSING FREQUENCIES

3.1 The chi-square test

Field biologists spend a good deal of their time counting and classifying things on nominal scales such as species, colour and habitat. Statistical techniques which analyse frequencies are therefore especially useful. The classical method of analysing frequencies is the chi-square test. This involves computing a test statistic which is compared with a chi-square (χ^2) distribution that we outlined in Section 11.11. Because there is a different distribution for every possible number of degrees of freedom (df), tables in Appendix 3 showing the distribution of χ^2 are restricted to the critical values at the significance levels we are interested in. There we give critical values at $P = 0.05$ and $P = 0.01$ (the 5% and 1% levels) for 1 to 30 df. Between 30 and 100 df, the critical values are estimated by interpolation, but the need to do this arises infrequently.

   Chi-square tests are variously referred to as tests for **homogeneity, randomness, association, independence** and **goodness of fit**. This array is not as alarming as it might seem at first sight. The precise applications will become clear as you study the examples. In each application the underlying principle is the same. The frequencies we *observe* are compared with those we *expect* on the basis of some Null Hypothesis. If the discrepancy between observed and expected frequencies is great, then the value of the calculated test statistic will exceed the critical value at the appropriate number of degrees of freedom. We are then obliged to reject the Null Hypothesis in favour of some alternative.

   The mastery of the method lies not so much in the computation of the test statistic itself but in the calculation of the expected frequencies. We have already shown some examples of how expected frequencies are generated. They can be derived from sample data (Example 7.5) or according to a mathematical model (Section 7.4). The versions of the test which compare observed frequencies with those expected from a model are called goodness of fit tests. All versions of the chi-square test assume that samples are random and observations are independent.

   Look again at the examples of χ^2 distribution in Fig. 11.3(b). The distribution is asymmetrical; a serious departure of *observed* from *expected* frequencies results in a χ^2 value in the right-hand tail of the distribution. But what do we make of a very small value of χ^2 that lies in the left-hand (truncated) tail? Very small values are as unlikely as very large values. Because statistical theory expects sampling error, or scatter, a small value of χ^2 means that observed frequencies agree with those expected so well as to be too good to be true! In practice, we are only interested in departures from agreement. In other words, if our Null Hypothesis H_0 is that observed and expected frequencies agree, then the alternative H_1 is that they do not. The test is therefore one-tailed and we ignore the possibility that they might agree too well.

13.2   Calculating the χ^2 test statistic

The simplest arithmetical comparison that can be made between an observed frequency and an expected frequency is the difference between them. In the test, the difference is squared and divided by the expected frequency. Thus:

$$\chi^2 = \frac{(O-E)^2}{E}$$

where O is an observed frequency and E is an expected frequency.

In practice, a series of observed frequencies is compared with corresponding expected frequencies resulting in several components of χ^2 all of which have to be summed. The general formula for χ^2 is therefore:

$$\chi^2 = \Sigma \frac{(O-E)^2}{E}$$

To illustrate the calculation of the χ^2 test statistic we choose a 'desk-top' example.

Example 13.1

A biologist writes a program in BASIC to produce random integers between 0 and 9 on a pocket computer. Such a program is:

```
10   PRINT INT(RAN#*10)
20 GOTO 10
```

After a few trials the biologist harbours a suspicion that the integers generated are not random. To test this, 100 numbers are generated and displayed as a frequency distribution:

Random number, x:	0	1	2	3	4	5	6	7	8	9	
Frequency of occurrence, f:			10	7	10	6	14	8	11	11	12	11	($n=100$)

The problem now is to work out how many times we should *expect* each integer to occur if they are indeed being drawn randomly.

This example permits two possible approaches to the problem, namely the *theoretical* and the *empirical*. First, independent integers generated randomly have, by definition, an equal probability of being drawn. Since the program allows only 10 integers, the probability of any one is 0.1. Using the relationship given in Section 7.3:

Expected frequency = probability of occurrence × sample size
$$= 0.1 \times 100$$
$$= 10$$

The expected frequency of each integer is 10. Alternatively, if our sample size, n, is 100, the Null Hypothesis predicts that there will be a *homogeneous* – that is, an evenly spread – distribution of numbers. Since there are 10 possible integers, a homogeneous distribution is $100/10 = 10$ of each. Thus, the expected frequency of each is again 10. The test can be either a test for *randomness* or a test for *homogeneity*, according to your point of view.

We can now write:

Random number, x:	0	1	2	3	4	5	6	7	8	9
Observed frequency, O:	10	7	10	6	14	8	11	11	12	11
Expected frequency, E:	10	10	10	10	10	10	10	10	10	10

Having worked out the expected frequencies, it is a simple matter to calculate the test statistic:

$$\chi^2 = \frac{(10-10)^2}{10} + \frac{(7-10)^2}{10} + \frac{(10-10)^2}{10} + \frac{(6-10)^2}{10} + \frac{(14-10)^2}{10} + \frac{(8-10)^2}{10}$$
$$+ \frac{(11-10)^2}{10} + \frac{(11-10)^2}{10} + \frac{(12-10)^2}{10} + \frac{(11-10)^2}{10}$$

$$\chi^2 = 0 + 0.9 + 0 + 1.6 + 1.6 + 0.4 + 0.1 + 0.1 + 0.4 + 0.1$$
$$\chi^2 = 5.2$$

Having calculated the test statistic, the next step is to evaluate the number of degrees of freedom (df). The rule is simple. It is the number of categories (a) less one. In this example there are therefore 9 df. We now have to determine whether the calculated value of 5.2 at 9 df exceeds the critical value at a chosen level of significance. This is decided by reference to Appendix 3. Turn to Appendix 3 and run down the left column until the number of degrees of freedom (9) is reached. Looking across the table at that point, we find two values: 16.92 under the 0.05 level of significance, and 21.67 under the 0.01. Our value of 5.2 is smaller than either of these. We accept the Null Hypothesis that the observed frequencies agree with the expected frequencies. We are justified in concluding that the integers *are* homogeneous and probably *were* generated independently and randomly. We can express the result of the test in the form: A statistical test shows that the frequency of each integer is consistent with having been generated randomly: $\chi_9^2 = 5.2$ (NS). The subscript 9 below the χ^2 is a shorthand way of indicating the 9 df. NS stands for *not significant*, that is, no significant difference between observed and expected frequencies.

13.3   A practical example of a test for homogeneous frequencies

Example 13.2

A biologist collects a sample of *Dixella* pupae from emergent vegetation in a pond. Different species cannot be identified when they are pupae; there is therefore no observer bias. It is assumed that larvae of all species are randomly dispersed as they pupate; in other words, individuals of all species are independent of each other and do not pupate in species aggregates. This is a fair assumption in the case of mobile aquatic larvae. On the basis of prior knowledge of the pond's community, the biologist prepares a table with four cells labelled *D. autumnalis*, *D. aestivalis*, *D. amphibia* and *D. attica*. Each cell represents a truly nominal category. As each adult midge emerges, it is identified and a tally-mark placed in the respective cell of the table. When emergence is complete the tallies are counted and the number in each category is expressed as a frequency. The results are:

D. autumnalis	D. aestivalis	D. amphibia	D. attica	Total
24	32	10	9	75

The question to be asked is: 'Is the distribution of frequencies between the categories homogeneous?' Alternatively: 'Can the variation between the observed frequencies be accounted for by chance scatter (sampling error), or is there really a significant difference between the frequencies?' In either case, the statistical hypotheses are:

H_0: The observed frequencies are homogeneous and the departure is merely due to sampling error or scatter.

H_1: The observed frequencies depart from those expected of a homogeneous distribution by an amount that cannot be explained by sampling error.

If H_0 is true we expect that the 75 pupae should be distributed homogeneously between the four nominal categories, that is $75/4 = 18.75$. The observed and expected frequencies are:

	D. autumnalis	D. aestivalis	D. amphibia	D. attica
O	24	32	10	9
E	18.75	18.75	18.75	18.75

Compute the test statistic exactly as in Example 13.1:

$$\chi^2 = \Sigma \frac{(O-E)^2}{E} = 1.47 + 9.36 + 4.08 + 5.07$$

$$= 19.98 \text{ with } (4-1) = 3 \text{ df}$$

From Appendix 3 we see that our calculated test statistic exceeds the critical values at both $P = 0.05$ and $P = 0.01$. We conclude that if the frequencies in the four categories *are* distributed homogeneously, then a discrepancy of this magnitude would happen by chance in fewer than 1 occasion in 100. We therefore reject H_0 in favour of H_1 and record the outcome, 'There is a statistically highly significant departure from homogeneity between the four categories, $\chi_3^2 = 19.98$, $P < 0.01$'. We may infer that there are significantly more *D. aestivalis* than other species, and that *D. attica* is in a minority.

13.4   The problem of independence

Objects which are dispersed randomly are independent of each other. Contagiousness (*aggregation*) is the result of attraction (*gregariousness*) and regularity the result of repulsion (*territoriality*), concepts that are explained in Chapters 7 and 8. Tests involving chi-square assume not only that samples are drawn at random but also that objects being counted are independent and randomly dispersed. These restrictions limit the number of useful questions that can be asked of a test.

Suppose the four frequencies in Example 13.2 had been the number of green-winged orchids counted in quadrats placed randomly in four different fields. Clearly, a chi-square test generates the same value of the test statistic, but how do we interpret the result? The only safe conclusion is that the counts are not homogeneous; the test does *not* tell us anything about differences between the fields. The data resemble those of Example 8.3. We can similarly obtain the variance of the four units (124.9) and divide it by the mean to obtain the dispersal index of $124.9/18.75 = 6.66$. Consult Fig. 8.2. Because the intersection of $\chi^2 (6.66 \times 3 \, \mathrm{df})$ and $v(3)$ is the above random zone, a contagious dispersal is accepted. Therefore, the orchids are not dispersed randomly and are not independent. The outcome of a homogeneity test tells us only that the orchids are not independent; it reveals nothing about differences between fields. *Only when we are sure that objects being counted are independent* can the test tell us anything about differences between fields. As we have said elsewhere, organisms are often *not* independent.

A further problem is that in the case of the quadrats, the items being counted are not sorted into nominal categories. An observation is, 'How many orchids are there in this quadrat?', rather than the question, 'Into which nominal category is this orchid to be placed?' A re-reading of the case of the midges in Example 13.2 will make this point explicit.

How then *should* we identify differences in the number of orchids between fields? The answer is to obtain *several* observations from each field, work out the average for each field and use a test for difference between averages (see Section 16.5).

13.5   One degree of freedom – Yates' correction

Where there are only two categories in a distribution there is only one degree of freedom. In this case, the calculated value of the test statistic is too high unless we make a correction called **Yates' Correction for Continuity**. This involves subtracting 0.5 from the numerator of each component of the chi-square formula before squaring. The subtraction is made from the *absolute value* of the difference $(O-E)$, that is, any minus sign is ignored. This is written as $(|O-E|-0.5)^2$ where the vertical bars on each side of $(O-E)$ mean *absolute value*.

Example 13.3

Sixteen chironomid (midge) larvae are found in a random benthic (bottom sediment) sample from a reservoir. The larvae are reared to adults which comprise 12 males and 4 females. Does this constitute a significant departure from a sex ratio of unity?

The Null Hypothesis H_0 is that the sex ratio *is* unity, that is, the ratio of males : females is 1 : 1. If this is the case we should expect 8 males and 8 females in the sample of 16. Sex is a truly nominal category. The χ^2 test for homogeneity is therefore appropriate. Because there is 1 df we apply Yates' correction:

- *without* the correction, the value of the test statistic is $[(12-8)^2/8 + (4-8)^2/8] = 4.0$. The value exceeds the tabulated critical value of 3.84 at 1df and is significant at $P=0.05$.
- *with* the correction it is $[(|12-8|-0.5)^2/8] + [(|4-8|-0.5)^2/8] = (1.531+1.531) = 3.062$. This is *less* than the tabulated critical value of 3.84 at $P=0.05$.

We conclude that there is *no* significant departure of the sex ratio from 1 : 1. Notice how in this particular example the application of the correction alters the outcome of the test.

We assume again that males and females are independently dispersed. This is a fair assumption in the case of mobile aquatic insect larvae but might not be so in the case of some mammals and birds that are known to aggregate into sex-biased groups in herds, roosts or hibernation sites. This possibility should be checked before undertaking the test.

13.6   Goodness of fit tests

The procedure for executing a **goodness of fit** test is exactly the same as for a homogeneity test except that the expected frequencies are generated according to a mathematical model. The rule for determining the degrees of freedom is different too.

NB

Expected frequency = probability of x sample size.

estimated

Example 13.4

Refer to Example 8.5. In Fig. 8.4 we compare the observed frequency distribution of nematodes in 60 grid squares (sampling units) in a counting chamber with the distribution predicted by a Poisson model. We say there, 'The agreement is very good, the small discrepancies being easily accounted for by random scatter'. We are now able to check that assertion. The observed and expected frequencies are:

	Frequency class x						
	0	1	2	3	4	5	6
O	3	12	17	13	9	3	1
E	4.458	11.58	15.06	13.08	8.46	4.44	1.92

Adding the seven components of $\Sigma(O-E)^2/E$ we obtain a test statistic of 1.66. In determining the number of degrees of freedom, we must bear in mind that the expected probability distribution from which the expected frequencies were estimated was based upon the estimation of a population parameter (λ) from a sample statistic. This facility 'costs' a degree of freedom *in addition* to the usual ($a-1$).

In this case, the degrees of freedom are therefore $(7-2)=5$. Our calculated test statistic is well below the critical value of 11.07 at $P=0.05$. Accordingly, H_0, that observed and expected frequencies are in agreement, is accepted. The distribution of nematodes is adequately described by the Poisson model.

Exactly the same procedure is followed in the case of binomial and negative binomial models. In both of these, however, *two* parameters are estimated from sample data (p, k and μ, k, respectively). In both cases then, the number of degrees of freedom is the number of frequency classes *a* minus 3.

13.7 Tests for association – the contingency table

In all of the previous examples of chi-square analysis, observed frequencies are distributed between categories in *one row*. When this is the case we refer to a **one-way classification**. Sometimes, however, two nominal level observations are obtained from a sampling unit. Thus, we may record an individual according to its age-class *and* sex; its species *and* habitat; and so on. In such cases, frequencies are arranged in *two or more rows* and we refer to a **two-way classification**. Tables of these data are called **contingency tables**. They allow the investigation of *association* between variables. The simplest type of contingency table is one which has only two nominal categories of each variable. It is called a **2 × 2 table**. An example follows.

Example 13.5

A biologist collects leaf litter from a 1 m^2 quadrat placed randomly at night on the ground in each of two woodlands – one on clay soil and the other on chalk soil. He sorts through the litter and picks out woodlice of two species, *Oniscus* and *Armadilidium*. It is assumed that the woodlice undertake their nocturnal foraging independently. The numbers of woodlice obtained are presented in the 2×2 contingency table (Table 13.1).

Cells in the table are conventionally labelled a, b, c, d and row totals, column totals and grand total are included.

Inspecting the distribution of frequencies in the table we see that there are more *Oniscus* than *Armadilidium* on clay but more *Armadilidium* than *Oniscus* on chalk. If a statistical test were to show that the proportional difference could not be accounted for by sampling error, then we would say they are significantly different. This would mean there is an *interaction* between the two variables. The interaction could be a positive association or a negative association. It appears that *Armadilidium* is positively associated with chalk and negatively associated with clay. If, however, a test were to show there is *no* significant difference between the proportions, then we would conclude that there is no interaction and the variables are independent. The test therefore, may be employed as a *test for association* or a *test for independence*, according to one's point of view. Do not confuse independent *variables* with independent *observations* (see Sections 2.5 and 13.4).

In order to apply the χ^2 test we need to calculate the expected frequency in each cell. This task is complicated by the fact we are in effect dealing with two Null Hypotheses:

(i) the ratios of the frequencies in both vertical columns do not depart from the overall vertical ratio (i.e. $14:22$ and $6:46$ do not depart from $20:68$)
(ii) the ratios of the frequencies in both horizontal rows do not depart from the overall horizontal ratio (i.e. $14:6$ and $22:46$ do not depart from $36:52$)

To calculate the expected frequency for any single cell we multiply the total for its column by the total for its row and divide the product by the grand total.

Table 13.1   Frequency of woodlice

	Oniscus	*Armadilidium*	*Totals*
Clay soil	**a** 14	**b** 6	**a+b** 20 ,
Chalk soil	**c** 22	**d** 46	**c+d** 68
Totals	**a+c** 36	**b+d** 52	**a+b+c+d** 88

GRAND TOTAL.

Thus, the expected frequency of *Oniscus* on clay is $[(a+c) \times (a+b)] \div (a+b+c+d) = 36 \times 20 \div 88 = 8.182$. The table may now be written out (in abbreviated form) showing the observed and expected frequencies in each cell.

Cell a	*Cell b*
$O = 14$	$O = 6$
$E = 36 \times 20 \div 88 = 8.182$	$E = 52 \times 20 \div 88 = 11.818$
Cell c	*Cell d*
$O = 22$	$O = 46$
$E = 36 \times 68 \div 88 = 27.818$	$E = 52 \times 68 \div 88 = 40.182$

(Check that the sum of the expected frequencies equals the sum of the observed frequencies, $a+b+c+d = 8.182 + 11.818 + 27.818 + 40.182 = 88$.)

With the four observed frequencies and their respective expected frequencies, the individual components of the test statistic may now be calculated. Before we do that, it is advisable to determine the number of degrees of freedom. The rule for determining the degrees of freedom in a contingency table is:

Degrees of freedom = number of columns (c) minus 1, multiplied by number of rows (r) minus 1

that is

$(c-1)(r-1)$

In our example this is $(2-1)(2-1) = 1$.

A 2×2 contingency table, therefore, has only one degree of freedom. With the conditions described in Section 13.5 in mind, we must apply Yates' correction to each cell in the table. Readers may note that a 2×2 table is the *only* one in which the correction needs to be applied.

Taking the four cells respectively, the test statistic is the sum of:

a	*b*
$(|14-8.182| -0.5)^2/8.182$	$(|6-11.818| -0.5)^2/11.818$
$= 3.457$	$= 2.393$
c	*d*
$(|22-27.818| -0.5)^2/27.818$	$(|46-40.182| -0.5)^2/40.182$
$= 1.017$	$= 0.704$

$$\chi^2 = 3.457 + 2.393 + 1.017 + 0.704$$
$$= 7.571$$

Consulting Appendix 3 we find that the calculated value of 7.571 at 1 df exceeds the tabulated critical value at both 5% and 1% levels of significance. There is therefore a statistically highly significant association between the variables. *Oniscus* is associated with clay soil and *Armadilidium* with chalk.

13.8   The $r \times c$ contingency table

Where there are more than two nominal categories in a two-way classification a contingency table has several rows and columns in it. If there are r rows and c columns there are $r \times c$ cells in the table. The procedure for working out the expected frequencies and calculating the test statistic is similar to that of a 2×2 table except that Yates' correction is not applied.

Example 13.6

Meniscus midges of the genus *Dixa* are found in running water and are indicators of surface pollutants (oil, hydrophobic chemicals, etc.). The mobile larvae pupate independently on emergent stones, vegetation, etc. where they may be collected. A biologist investigating habitat preferences of four species of *Dixa* obtains randomly pupae from three streams of different degrees of eutrophication (oligotrophic, mesotrophic and eutrophic). The frequencies obtained are shown in Table 13.2.

All categories in both directions are nominal; chi-square contingency analysis is therefore appropriate. The steps in the analysis are as follows:

(i) Calculate the 12 individual frequencies that would be expected if H_0 is true, that is, there is no association between any of the categories. By way of example, the expected frequency of *D. nebulosa* in oligotrophic is $(61 \times 41) \div 157 = 15.93$ (column total $\times$ row total $\div$ grand total).

(ii) Calculate all the individual components of χ^2 from $(O - E)^2 / E$. The results of the procedure thus far are tabulated in Table 13.3.

(iii) Sum all the individual values of $(O - E)^2 / E$. The result is 30.983. This is the test statistic.

(iv) Determine the degrees of freedom from $(c - 1)(r - 1)$. This is $3 \times 2 = 6$.

(v) Consult Appendix 3 at this number of degrees of freedom and decide if the test statistic calculated in step (iii) exceeds the critical value at $P = 0.05$ or $P = 0.01$. The critical value at $P = 0.01$ is 16.81. The calculated value exceeds this.

(vi) Express the result in the form: 'There is a statistically highly significant association between certain *Dixa* species and state of eutrophication; $\chi_6^2 = 30.98$, $P < 0.01$'.

In order to decide which *Dixa* species are associated with which eutrophication state, we examine the individual components of χ^2 in Table 13.3. Inspecting the table, we find that the largest individual values are: 5.577 (*D. nubilipennis* in oligotrophic); 10.88 (*D. dilatata* in mesotrophic); 3.765 (*D. nebulosa* in eutrophic); and 3.56 (*D. dilatata* in eutrophic). Looking at those cells of the table more closely, there are more *D. nubilipennis* than expected in oligotrophic; more *D. dilatata* than expected in mesotrophic; and more *D. nebulosa* than expected in eutrophic. We conclude that these species are positively associated with the respective conditions. However, there are fewer *D. dilatata* than expected in eutrophic and we infer that there is a negative association between this species and this condition.

Table 13.2 Frequencies of *Dixa* species

Site	Species				
	D. nebulosa	*D. submaculata*	*D. dilatata*	*D. nubilipennis*	*Totals*
Site 1: Oligotrophic	12	7	5	17	41
Site 2: Mesotrophic	14	6	22	9	51
Site 3: Eutrophic	35	12	7	11	65
Totals	61	25	34	37	157

Table 13.3 Calculation of expected frequencies

Site	Variable a: species			
	D. nebulosa	D. submaculata	D. dilatata	D. nubilipennis
Site 1: Oligotrophic				
O:	12	7	5	17
E:	15.93	6.53	8.88	9.66
$(O-E)^2/E$:	0.970	0.034	1.695	5.577
Site 2: Mesotrophic				
O:	14	6	22	9
E:	19.82	8.12	11.04	12.02
$(O-E)^2/E$:	1.709	0.553	10.88	0.759
Site 3: Eutrophic				
O:	35	12	7	11
E:	25.25	10.35	14.08	15.32
$(O-E)^2/E$:	3.765	0.263	3.56	1.218

13.9 The *G*-test

The **G-test** is an alternative to the chi-square test for analysing frequencies. The two methods are interchangeable; if a chi-square test is appropriate then so too is a *G*-test and the assumptions in each are the same. Moreover, the outcome of the *G*-test is a test statistic (*G*) which is compared with the distribution of chi-square in the same tables as the chi-square test. Why then do we need a second test that serves exactly the same purpose?

First, the *G*-test is easier to execute with a desk-top hand calculator, especially with contingency tables. Second, mathematicians believe that the *G*-test has theoretical advantages in advanced applications which are beyond the scope of this text. Despite these advantages, however, it seems that 'old habits die hard'; chi-square is by far the most widely used method of analysing frequencies in current ecological journals.

13.10 Applying the *G*-test to a one-way classification of frequencies

The *G*-test may be used for testing *homogeneity* and for *goodness of fit* of frequencies arranged in a one-way classification. The procedure in each is the same, remembering that expected frequencies in a goodness of fit test are generated according to a mathematical model and the rule for determining the degrees of freedom is different. A correction factor (**Williams' correction**) is applied in the *G*-test – irrespective of the number of degrees of freedom. (But see

Section 13.12 and the case of a $r \times c$ contingency table.) The adjusted value of G is symbolized G_{adj}.

To emphasize the similarity between the G-test and the chi-square test, we use the data employed in Example 13.2. Observed and expected frequencies are:

O: 24   32   10   9
E: 18.75   18.75   18.75   18.75

The formula for G is:

$$G = 2 \times \sum^{a} O \ln \frac{O}{E}$$

where O and E are *observed* and *expected* frequencies, respectively; $\sum^{a}$ means the sum of the products $O \ln(O/E)$ for all a categories or frequency classes; and 'ln' means natural logarithm. The stepwise procedure is as follows:

Step 1

For each category or frequency class multiply O by $\ln(O/E)$ and add up the total. Thus:

$$24 \times \ln\left(\frac{24}{18.75}\right) + 32 \times \ln\left(\frac{32}{18.75}\right) + 10 \times \ln\left(\frac{10}{18.75}\right) + 9 \times \ln\left(\frac{9}{18.75}\right)$$

$$= 5.925 + 17.105 + (-6.286) + (-6.606)$$
$$= 10.138$$

Step 2

Double this number:

$$10.138 \times 2 = 20.276$$

This is the test statistic, G

Step 3

Divide G by a correction factor which is applied irrespective of the number of degrees of freedom.

Correction factor $= 1 + (a^2 - 1)/6nv$

where a is the number of categories or frequency classes, n is the total number of observed frequencies and v is the degrees of freedom $(a-1)$. Thus

Correction factor $= 1 + (4^2 - 1)/(6 \times 75 \times 3)$
$$= 1 + (15/1350)$$
$$= 1 + 0.0111 = 1.0111$$
$G_{adj} = G/\text{correction factor} = 20.276/1.0111 = 20.053$

Step 4
Compare the value of G_{adj} against the chi-square distribution (Appendix 3) for $(a-1)=3$ df.

The calculated value of G_{adj} is very similar to the value of 19.98 obtained in the chi-square test in Example 13.2.

There is no difference in the execution of the *G*-test when there is only one degree of freedom. Readers may apply the *G*-test to the data in Example 13.3 and confirm that $G=4.186$, the correction factor is 1.031 and G_{adj} is $4.186/1.031=4.06$.

In using the *G*-test for goodness of fit, the rules for determining the degrees of freedom are the same as in the chi-square test (Section 13.6).

13.11   Applying the *G*-test to a 2×2 contingency table

In using the *G*-test for analysing 2×2 and $r \times c$ contingency tables we do not have to distinguish between *observed* and *expected* frequencies. We symbolize the observed frequencies therefore simply as *f*. Observed frequencies are multiplied by the natural logarithm of themselves and the products summed ($\Sigma f.\ln f$). This operation is easily accomplished with a scientific calculator.

Example 13.7

Obtain $\Sigma f.\ln f$ for the frequencies 5 and 8.

Operation	Readout
⑤ × ln M+	8.047189562
⑧ × ln M+.	16.63553233
MR	24.68272189

Thus, $(5.\ln 5)+(8.\ln 8)=24.683$ to three decimal places. Note that the operation M+ executes both the multiplication of *f* by ln*f* and adds the product to the sum accumulated in the memory.

The procedure for using the *G*-test with a 2×2 contingency table is shown in Example 13.8.

Example 13.8

A trawl sample from the North Sea contained 86 flounders without skin lesions and 15 with lesions. A trawl sample from the Irish Sea contained 32 flounders without lesions and 12 with lesions. Is there a statistically significant difference between the two proportions?

The steps in the execution of the *G*-test are:

Step 1

Display the observed frequencies in a contingency table showing row, column and grand totals.

	Without lesions	*With lesions*	*Totals*
North Sea	**a** 86	**b** 15	**a + b** 101
Irish Sea	**c** 32	**d** 12	**c + d** 44
Totals	**a + c** 118	**b + d** 27	**a + b + c + d** 145

Step 2

Calculate $\Sigma f.\ln f$ for all observed frequencies:

$(86.\ln 86) + (15.\ln 15) + (32.\ln 32) + (12.\ln 12)$
$= 564.417$

Step 3

Calculate $f.\ln f$ for the grand total $(a + b + c + d)$:

$145.\ln 145 = 721.626$

Step 4

Calculate $\Sigma f.\ln f$ for all row and column totals:

$(101.\ln 101) + (44.\ln 44) + (118.\ln 118) + (27.\ln 27)$
$= 1284.560$

Step 5

Add the numbers obtained in Steps 2 and 3 and subtract the number obtained in Step 4:

Step 2 + Step 3 − Step 4
$= 564.417 + 721.626 - 1284.560$
$= 1.483$

Step 6

Double this number to give G:

$G = 1.483 \times 2 = 2.966$

Step 7

Calculate Williams' correction factor from:

$$\text{Correction factor} = 1 + \frac{\left[\left(\frac{n}{a+b}\right) + \left(\frac{n}{c+d}\right) - 1\right]\left[\left(\frac{n}{a+c}\right) + \left(\frac{n}{b+d}\right) - 1\right]}{6n}$$

$$= 1 + \frac{\left[\left(\frac{145}{101}\right) + \left(\frac{145}{44}\right) - 1\right]\left[\left(\frac{145}{118}\right) + \left(\frac{145}{27}\right) - 1\right]}{6 \times 145}$$

$$= 1 + \frac{20.891}{870}$$

$$= 1 + 0.024 = 1.024$$

Step 8

Divide G by the correction factor to obtain G_{adj}:

$$G_{adj} = 2.966/1.024 = 2.896$$

Step 9

Compare G_{adj} with chi-square in Appendix 3 at $(r-1)(c-1) = 1$ df. The value is below the critical value of 3.84 at $P = 0.05$. There is no significant difference between the proportions in the North Sea and the Irish Sea.

Comparing the value of G_{adj} with that of the calculated test statistic in the chi-square test (with Yates' correction applied), the latter is 2.355, lower than the value of G_{adj}.

It is, of course, very unlikely that individual flounders are independent – they aggregate into loose shoals. However, provided that the two *categories* are independent of each other and lesioned fish mix randomly in the population as a whole and do not isolate themselves into separate assemblages, the chi-square or G-test may be validly applied.

13.12   Applying the *G*-test to an $r \times c$ contingency table

The procedure for applying the G-test to an $r \times c$ table is exactly the same as for a 2×2 table. However, the correction factor is not usually applied because it is so small that it may be ignored.

Example 13.9

A biologist investigating the population dynamics of feather lice (Mallophaga) obtains samples from storm petrels in July and September. Lice are sorted into age-classes corresponding to 1st, 2nd, 3rd instar nymphs and adults. The frequencies are, respectively, 38, 70, 32, 70 in July and 4, 22, 11, 42 in

September. Is there a statistically significant difference between the proportions of the age-classes in the two samples?

The steps in the execution of the *G*-test are as follows:

Step 1
Display the observed frequencies in a contingency table showing row, column and grand totals.

	1st instar nymphs	2nd instar nymphs	3rd instar nymphs	Adults	Totals
July	38	70	32	70	210
September	4	22	11	42	79
Totals	42	92	43	112	289

Step 2
Calculate $\Sigma f.\ln f$ for all observed frequencies:

$(38.\ln 38) + (70.\ln 70) + (32.\ln 32) + (70.\ln 70) + (4.\ln 4) + (22.\ln 22) + (11.\ln 11) + (42.\ln 42)$
$= 138.228 + 297.395 + 110.904 + 297.395 + 5.545 + 68.003 + 26.377 + 156.982$
$= 1100.829$

Step 3
Calculate $f.\ln f$ for the grand total:

$289.\ln 289 = 1637.597$

Step 4
Calculate $\Sigma f.\ln f$ for all row and column totals:

$(210.\ln 210) + (79.\ln 79) + (42.\ln 42) + (92.\ln 92) + (43.\ln 43) + (112.\ln 112)$
$= 1122.893 + 345.186 + 156.982 + 416.005 + 161.732 + 528.472$
$= 2731.27$

Step 5
Calculate Step 2 + Step 3 − Step 4

$1100.829 + 1637.597 - 2731.27$
$= 7.156$

Step 6
Double this to give *G*:

$7.156 \times 2 = 14.312$

Step 7

No correction is applied. Compare G with the chi-square distribution in Appendix 3 at $(r-1)(c-1) = 1 \times 3 = 3$ df. The calculated G is larger than the critical value of 11.34 at $P = 0.01$. We therefore conclude that the proportions are statistically highly significantly different. The corresponding value of the chi-square test statistic is 13.143, smaller than the calculated G.

The biologist may infer that the proportion of adults in the later sample has increased at the expense of the lower instar nymphs, suggesting that the population is ageing and reproductive rates of the lice are declining in September.

13.13   Advice on analysing frequencies

1. All versions of the chi-square test compare the agreement between a set of observed frequencies and those expected if some Null Hypothesis is true. They are all, in a sense, *goodness of fit* tests, although this title is usually restricted to those in which the expected frequencies are estimated according to a mathematical model or, in genetics, by phenotypic frequencies predicted by some ratio. *G*-tests are similar. But note that in contingency tables the intermediate estimation of expected frequencies is not required in the *G*-test.

2. As objects are counted they should be assigned to nominal categories. Example 13.2 emphasizes this point. Unambiguous intervals on a continuous scale may be regarded as *nominal* for the application of the tests. For example, pH bands of 4.1–6.0, 6.1–8.0, 8.1–10.0 represents three categories into which frequencies can be assigned. Our *state of eutrophication* in Example 13.6 and *month* in Example 13.9 are categories of this kind.

3. The application of the tests requires that samples are random and objects counted are independent and therefore randomly dispersed. If they are not, then rejection of H_0 is more likely to be the result of the aggregation of items, that is, their lack of independence rather than the effect of the variable in question. The requirement of independence is often a major restriction in analysing frequencies and steps must be taken to check this. The caveat, '*it is assumed that observations are independent*', may not always be convincing!

4. The sample size, that is, the grand total of observed frequencies (n), should be such that all *expected* frequencies exceed 5. In marginal cases this can sometimes be achieved by collapsing cells and aggregating the respective observed frequencies and expected frequencies. Some flexibility in interpreting this rule is allowed. Most statisticians would not object to some of the expected frequencies being below 5, provided that no more than one-fifth of the total number of expected frequencies is below 5, and none are below 1.

5. All versions of the tests require that observed frequencies are, indeed, *actually observed*. They must not be estimates or derived variables. For example, in Example 13.8 the flounders with lesions could be expressed as

14.85% in the North Sea and 27.27% in the Irish Sea. Without the actual number of fish counted, however, we are unable to test the difference between the proportions.

6. Apply Yates' correction in the chi-square test when there is only one degree of freedom. In the G-test apply Williams' correction in all one-way classification tests and in 2×2 contingency tables. It may be ignored in larger contingency tables.

7. The problem might arise whether to choose the chi-square or the G-test. They are totally interchangeable and if one is applicable, then so too is the other. The underlying assumptions of each are the same. In analysing contingency tables the G-test has the distinct advantage that expected frequencies are not separately estimated and the test is quicker to execute using a scientific calculator.

The calculated values of χ^2 and G (or G_{adj}) for a given set of data are similar. It is only in marginal cases where the test statistic is close to the tabulated critical value that it matters which test you use. G is invariably a little larger than χ^2 and so the G-test tends to reject H_0 more often than chi-square. You should *not* exploit this fact to obtain a desirable outcome of a test – that is cheating! Neither should you do one test and then the other to see how they compare. Which test you are to employ is decided at the outset of a project or research programme and then used throughout.

In the biological literature the chi-square test is still the favourite but the G-test appears to be gaining in popularity as its flexibility is recognized. For further information on the application of the G-test we suggest you consult Sokal and Rohlf (1981).

14 MEASURING CORRELATIONS

14.1 The meaning of correlation

Many variables in nature are related; examples from biology include the mass of a growing organism and its volume, the length of an otolith ('ear-stone') and the length of the fish it is taken from, the structural complexity of a plant community and latitude. Relationships or associations between variables such as these are referred to as **correlations**. Correlations are measured on ordinal or interval scales.

When an increase in one variable is accompanied by an increase in another, the correlation is said to be **positive** or **direct**. The length of an otolith and the length of the fish are positively correlated. When an increase in one variable is accompanied by a decrease in another, the correlation is said to be **negative** or **inverse**. The mass of body fat of a migrating bird and the distance flown since its last feed are negatively correlated.

The fact that variables are associated or correlated does not necessarily mean that one *causes* the other. Otolith length and body length in a population of fish may be correlated but one cannot be said to cause the other; both are undoubtedly related to some underlying genetic factor. In common usage, the word 'correlation' describes any type of relationship between objects and events. In statistics however, correlation has a precise meaning; it refers to a quantitative relationship between two variables measured on ordinal or interval scales.

14.2 Investigating correlation

Bivariate observations of variables measured on ordinal or interval scales can be displayed as a scattergram (Figs 4.8 and 14.1). Just as a simple dot-diagram gives both a useful indication of whether a sample of observations is roughly symmetrically distributed about a mean and the extent of the variability, a scattergram gives an impression of correlation. Figure 14.1(a) shows a clear case of a positive correlation, whilst Fig. 14.1(b) shows an equally clear case of a negative correlation. Figure 14.1(c) shows no correlation, but what about Fig. 14.1(d)? It is not easy to be certain about this.

Subjective examination of a scattergram must be replaced by a statistical

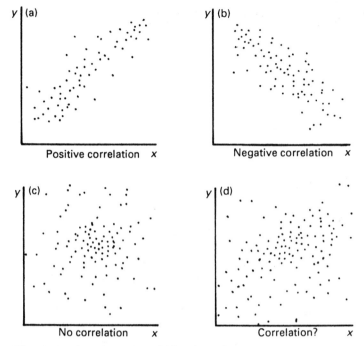

Fig. 14.1 Scattergrams of bivariate data.

technique which is more precise and objective. The statistic that provides an index of the degree to which the two variables are related is called the **correlation coefficient**. The statistic is calculated from sample data and is the estimator of the corresponding population parameter.

The numerical value of the correlation coefficient, r, falls between two extreme values: $+1$ (for perfect positive correlation) and -1 (for perfect negative correlation). A perfect correlation exists when all the points in a scattergram fall on a perfectly straight line, as indicated in Fig. 14.2.

Perfect or near perfect correlations (positive or negative) are virtually non-existent in biological situations; they are the privilege of the physicist! An

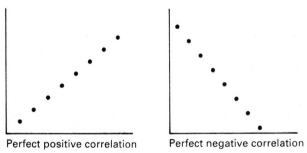

Fig. 14.2 Perfect correlations.

example of a perfect correlation is described in Section 15.4. A correlation coefficient of 0, or near 0, indicates lack of correlation.

Correlation coefficients can be calculated by both non-parametric and parametric methods. Of the various non-parametric coefficients, the **Spearman Rank Correlation Coefficient** is among the most widely used and is the one we illustrate. It is appropriate to observations based on ordinal as well as interval scales. A parametric coefficient is the **Product Moment Correlation Coefficient**. It is used only for interval scale observations and is subject to more stringent conditions than non-parametric alternatives.

14.3 The strength and significance of a correlation

In the previous section we said that all values of a correlation coefficient fall on a scale with limits -1 to $+1$. The closer the value of a coefficient is to -1 or $+1$, the greater is the strength of the correlation, whilst the closer it is to 0 the weaker it is. As a rough and ready guide to the meaning of the coefficient Table 14.1 offers a descriptive interpretation.

Sometimes an apparently strong correlation may be regarded as *not statistically significant* whilst a weak correlation may be *statistically highly significant*. We must resolve this apparent paradox. Look again at Fig. 14.1(c) which we say shows no correlation. Imagine that the points in the scattergram represent a population of observations from which individual points can be drawn at random. The first two points drawn are bound to be in perfect alignment because any two points can be connected by a straight line. We should not be tempted however into thinking in terms of correlation! It is not improbable that a third point, drawn randomly, could be in rough alignment with the first two, giving an impression of correlation. However, the more points that are drawn the less and less likely it becomes that a chance correlation can be maintained. When *all* the points in the population have been drawn, we can measure the absolute correlation coefficient *parameter* ρ (rho). Any sample drawn from the population estimates ρ. Large samples give reliable estimates and small samples give less reliable estimates. If the value of ρ is low, that is, a weak correlation in the population, large samples give good estimates and are *statistically significant*. On the other hand, if ρ should be large, a small

Table 14.1 The strength of a correlation

Value of coefficient r (positive or negative)	Meaning
0.00 to 0.19	A very weak correlation
0.20 to 0.39	A weak correlation
0.40 to 0.69	A modest correlation
0.70 to 0.89	A strong correlation
0.90 to 1.00	A very strong correlation

sample is a poor estimate which may be not statistically significant. The point to bear in mind is that larger samples do not *strengthen* a weak correlation. But they do reduce the likelihood of a spurious correlation arising by chance or sampling error.

4.4 The Spearman Rank Correlation Coefficient r_S

The formula for calculating r_S is:

$$r_S = 1 - \left[\frac{6\Sigma d^2}{(n^3 - n)} \right]$$

where n is the number of units in a sample, d is the difference between ranks, Σ is the 'sum of' and 6 is a constant peculiar to this formula.

Because observations are converted to *ranks* the test may be applied to observations made on *ordinal* scales. It is suitable for samples with between 7 and 30 pairs of observations. The following example shows how the formula is applied.

Example 14.1

A county naturalists' trust surveys a sample of meadow habitats around the county and obtains measures of the diversity of grasshoppers collected in standardized sweep nets (see Section 11.10). For those meadows of which the management history is known, the number of years elapsed since pesticides were applied is recorded. The observations are set out in Table 14.2.

Table 14.2 Survey of meadow habitats

Variable x: Years since pesticides applied	Rank of x	Variable y: Diversity of grasshoppers H	Rank of y	d	d²
0	1	0	1	0	0
1	2	0.19	3	−1	1
3	3	0.15	2	1	1
5	4	1.49	8	−4	16
(8–10)	5	1.10	4	1	1
12	6	1.12	5	1	1
13	7	1.61	9	−2	4
15	8	1.42	6	2	4
21	9	1.48	7	2	4
25+	10	1.92	10	0	0
					$\Sigma d^2 = 32$

The x variable is the number of years elapsed. Accurate records are not available in all cases; the 5th meadow in the list was last treated between 8 and 10 years ago; the last meadow has no recorded history of pesticide use. Though all observations are not precise, they can be listed in rank order with the ranks shown in the next column. The y variable is the diversity index H recorded for the corresponding meadows listed in the x column. Although H is a continuous derived variable, it is not measured on a linear interval scale; '2' does *not* mean twice as diverse as '1'. However, H is ranked numerically on an ordinal scale and the ranks of each observation are listed in the 4th column.

The actual values of the observations are now discarded and the *ranks* become the basic data used in the test. In the column headed d is the arithmetic difference between the pairs of ranks. When the figures in this column are totalled they provide a *sum of differences*. Since this always equals zero it is not a helpful statistic. However, it should always be checked. If it does not equal zero, then there is an error in the ranking. Of more use is the sum of the squares of the differences, Σd^2. This is obtained by squaring each of the values of d and adding them up. Σd^2 is used in the formula for calculating r_S.

Substituting in the formula for r_S:

$$r_S = 1 - \left[\frac{(6 \times 32)}{(10 \times 10 \times 10) - 10} \right] = 1 - \left(\frac{192}{990} \right) = (1 - 0.194) = 0.806$$

A value of 0.806 suggests a strong positive correlation according to Table 14.1. However, we must first check that such a value could not have arisen by chance in a sample of 10 units if, in fact, there is no correlation within the population. Our hypotheses are:

H_0: There is no correlation: the value of r_S is obtained by chance (sampling error).

H_1: A value of r_S as great as 0.806 could not be the result of sampling error in a sample size of 10 units.

The value of r_S is compared with the distribution shown in Appendix 4. Entering the table at $n = 10$, we see that our calculated value exceeds the critical tabulated value of 0.794 at $P = 0.01$. We reject H_0 and say the correlation is highly significant. We may infer that the diversity of grasshoppers in meadows is strongly positively correlated with time elapsed since pesticides were last used.

Our second example of the Spearman Rank Correlation Coefficient involves a negative correlation in which some of the observations are equal in value. These are dealt with as explained in Section 3.6.

Example 14.2

A biologist investigates the usefulness of Plecoptera (stonefly) nymphs as indicators of environmental factors in streams. Samples from 13 streams are obtained by displacing nymphs from a stream bed into a net by means of a standardized-kick technique. Values of water hardness – calcium carbonate concentration – are obtained from the local water authority. The observations and their ranks, together with d and d^2 are shown in Table 14.3. Is there a significant correlation between water hardness and number of Plecoptera nymphs?

Table 14.3   Stonefly nymphs and water hardness

Variable x: Water hardness ($CaCO_3$ units)	Rank of x	Variable y: Number of Plecoptera nymphs	Rank of y	d	d^2
17	1	42	13	−12	144
20	2	40	12	−10	100
22	3	30	11	−8	64
28	4	7	6	−2	4
42	5	12	10	−5	25
55	$6\frac{1}{2}$	10	9	−2.5	6.25
55	$6\frac{1}{2}$	8	8	−1.5	2.25
75	8	7	6	2	4
80	9	3	2	7	49
90	10	7	6	4	16
145	$11\frac{1}{2}$	5	4	7.5	56.25
145	$11\frac{1}{2}$	2	1	10.5	110.25
170	13	4	3	10	100
					$\Sigma d^2 = 681$

Within the x variable the observations 55 and 145 occur twice each; they are given the average ranks of $6\frac{1}{2}$ and $11\frac{1}{2}$, respectively. In the y variable the observation 7 occurs three times and is given the average rank of 6.
   Substituting in the formula:

$$r_s = 1 - \left[\frac{(6 \times 681)}{(13 \times 13 \times 13) - 13} \right]$$

$$= 1 - \left(\frac{4086}{2184} \right)$$

$$= (1 - 1.87) = -0.87$$

The correlation coefficient appears strong and negative. Checking the significance in Appendix 4 (ignore the minus sign for this purpose) we find that the value of 0.87 exceeds the tabulated critical value of 0.745 for $n = 13$ at $P = 0.01$. We reject H_0 and conclude that the inverse correlation is statistically highly significant.
   It must be emphasized that the strong inverse correlation does not establish that hard water is the *cause* of low numbers of stonefly nymphs in streams. It is

possible that another, as yet unidentified, factor is the underlying source of variation in both variables. However, because the correlation is significant, the biologist may infer that streams with low numbers of stonefly nymphs are indicative of hard water, at least within the general area from which the initial samples were obtained.

An advantage of the Spearman Rank Correlation Coefficient is that it may be applied to observations measured on ordinal scales. Moreover, there is no requirement that data are normally distributed; that is to say, it is *distribution free*. That is why we can apply it to the numbers of Plecoptera nymphs in Example 14.2. The variance to mean ratio (197 : 13.6) in the sample of nymphs is indicative of a highly skewed distribution which is far from normal (see Example 8.3).

When observations are measured on interval scales and are approximately normally distributed, then a parametric correlation coefficient is appropriate.

14.5 Covariance

Listed below are four pairs of bivariate observations of variables x and y measured on interval scales.

x units	y units
4	10
4	6
8	10
8	6

Number of pairs of observations, $n = 4$
Mean of the x observations, $\bar{x}$, $\Sigma x/4 = 6$
Mean of the y observations, $\bar{y}$, $\Sigma y/4 = 8$

The observations are displayed in Fig. 14.3 as a scattergram. The scattergram is divided into four quadrants by dashed lines through $\bar{x}$ and $\bar{y}$. The intersection at $\bar{x}, \bar{y}$ (6,8) is called the **mean centre**.

Each point can be described in terms of its deviation from each dashed line, that is to say, the deviation from $\bar{x}$ and the deviation from $\bar{y}$. The individual deviations are shown as dotted lines. In our example all deviations are 2 units. Deviations above the $\bar{y}$ line and to the right of the $\bar{x}$ line are positive; deviations below the $\bar{y}$ line and to the left of the $\bar{x}$ line are negative.

When the two deviations of a point are multiplied together, the product is called the **product of deviations**.

- Products of deviations in quadrant A are negative
- Products of deviations in quadrant B are positive
- Products of deviations in quadrant C are positive
- Products of deviations in quadrant D are negative

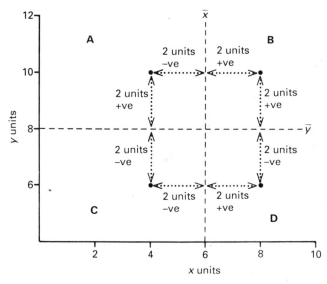

Fig. 14.3 Deviations from the mean centre.

Summing the products of deviations for all points gives the **sum of products**:

Quadrant	Deviation from $\bar{x}$		Deviation from $\bar{y}$	Product of deviations
A	−2	×	+2=	−4
B	+2	×	+2=	+4
C	−2	×	−2=	+4
D	+2	×	−2=	−4
			Sum of products	=0

Because of the symmetrical disposition of the points in the scattergram, the positive products of deviations and the negative products of deviations cancel out, giving a sum of products that is zero. Look again at Fig. 14.1 drawn again as Fig. 14.4.

In Fig. 14.4(a) most of the points are in quadrants C and B, with very few in A and D. The positive products of deviations outnumber the negative products of deviations and the sum of products therefore has a net positive value. In Fig. 14.4(b) the converse is true; most of the points are in quadrants A and D resulting in a sum of products with a negative value. In both these cases there is correlation, positive and negative, respectively. In Fig. 14.4. (c) there are similar numbers of points in all four quadrants and so the negative products of deviations roughly cancel out the positive products of deviations and the sum

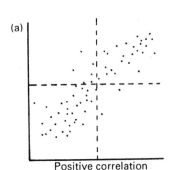

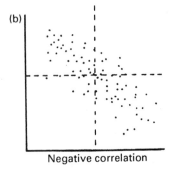

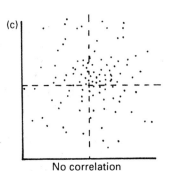

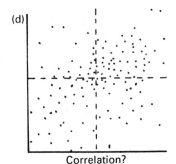

Fig. 14.4   The idea of covariance.

of products is near to zero. In Fig. 14.4(d) the value of the sum of products can help us to decide whether or not there is a correlation.

The sum of products of a set of bivariate data is calculated from:

$$\Sigma(x-\bar{x})(y-\bar{y})$$

When this is divided by the degrees of freedom (number of pairs of observations minus 1) we obtain an important measure of correlation called the **covariance**:

$$\text{Covar}_{x,y} = \frac{\Sigma(x-\bar{x})(y-\bar{y})}{(n-1)}$$

14.6   The Product Moment Correlation Coefficient

The disadvantage of using covariance as a measure of correlation between x and y is that its numerical value depends on the units in which x and y are measured. This makes comparisons difficult. However, when the covariance is divided by the standard deviation of x and the standard deviation of y it is standardized on to a scale with absolute limits of -1 and $+1$ and is called the **Product Moment Correlation Coefficient**, r.

$$r = \frac{\dfrac{\Sigma(x-\bar{x})(y-\bar{y})}{(n-1)}}{s_x \times s_y}$$

In practice, an algebraic rearrangement of the formula is easier to use:

$$r = \frac{n\Sigma xy - \Sigma x \Sigma y}{\sqrt{[n\Sigma x^2 - (\Sigma x)^2][n\Sigma y^2 - (\Sigma y)^2]}}$$

How to apply this formula is shown in Example 14.3.

Example 14.3

Many species of fish contain in their heads a pair of characteristic 'ear-stones' called otoliths. These resist absorption in digestive tracts and may often be found in seal droppings and seabird regurgitates. If a good correlation were found to exist between the length of otoliths and a variable such as the mass or the length of the fish from which they came, then biologists would be able to draw inferences about the diets of seals or seabirds from otoliths found in droppings or regurgitates.

A sample of 10 fish of one species is randomly selected from a box on a fishing boat. Each fish is weighed and measured and then dissected to remove the otoliths, which are also measured. It is assumed that these variables are normally distributed within the population from which the sample is drawn. The data are tabulated in Table 14.4 as a set of bivariate data. The otolith measurements may be regarded as the mean of the pair obtained from each fish. Columns are provided for x^2, y^2 and xy. To obtain xy multiply each x value by its matched y value. The sums of all values (Σ) are given at the foot of Table 14.4.

Table 14.4 Otolith and fish length measurements

Otolith length x (mm)	Fish mass y (g)	x^2	y^2	xy
6.6	86	43.56	7 396	567.6
6.9	92	47.61	8 464	634.8
7.3	71	53.29	5 041	518.3
7.5	74	56.25	5 476	555
8.2	185	67.24	34 225	1 517
8.3	85	68.89	7 225	705.5
9.1	201	82.81	40 401	1 829.1
9.2	283	84.64	80 089	2 603.6
9.4	255	88.36	65 025	2 397
10.2	222	104.04	49 284	2 264.4
82.7	1554	696.69	302 626	13 592.3

Summarizing the data from Table 14.4:

$\Sigma x = 82.7$	$\Sigma y = 1554$	$n = 10$
$(\Sigma x)^2 = 82.7^2$	$(\Sigma y)^2 = 1554^2$	$\Sigma xy = 13\,592.3$
$= 6839.29$	$= 2\,414\,916$	
$\Sigma x^2 = 696.69$	$\Sigma y^2 = 302\,626$	

Substituting in the formula for the product moment correlation coefficient:

$$r = \frac{(10 \times 13\,592.3) - (82.7 \times 1554)}{\sqrt{[(10 \times 696.69) - 6839.29)][(10 \times 302\,626) - 2\,414\,916]}}$$

$$r = \frac{(135\,923) - (128\,515.8)}{\sqrt{[127.61] \times [611\,344]}} = \frac{7407.2}{\sqrt{78\,013\,607.84}} = \frac{7407.2}{8832.53}$$

$$r = 0.8386$$

According to Table 14.1 there appears to be a strong positive correlation between otolith length and mass of fish. We need to check that such a correlation is unlikely to have arisen by chance in a sample of 10 units. Appendix 5 gives the probability distribution of r. First we need to work out the degrees of freedom. In the case of a correlation coefficient they are the number of pairs of observation less 2, that is, $(n-2) = 8$ in our example. Consulting Appendix 5 we find that our calculated value of 0.8386 exceeds the tabulated value at 8 df of 0.765 at $P = 0.01$. Our correlation is therefore statistically highly significant.

Example 14.4

Replace the observations of fish mass in Example 14.3 (Table 14.4) by 10 observations of fish *length* (cm). In respective order the observations are:

16.4   17.6   18.4   18.2   19.9   20.1   22.3   22.3   24.4   26.5

Follow the procedure outlined in Example 14.3 to confirm the following data and then calculate the correlation between otolith length and fish length.

$\Sigma x = 82.7$	$\Sigma y = 206.1$	$n = 10$
$(\Sigma x)^2 = 82.7^2$	$(\Sigma y)^2 = 206.1^2$	$\Sigma xy = 1738.26$
$= 6839.29$	$= 42\,477.21$	
$\Sigma x^2 = 696.69$	$\Sigma y^2 = 4340.73$	$r = 0.9815$ with 8 df

[*Hint*: many of the terms may be obtained directly on a scientific calculator – see Section 6.8.]

14.7   The coefficient of determination r^2

The square of the product moment correlation coefficient is itself a useful statistic and is called the **coefficient of determination**. It is a measure of the proportion of the variability in one variable that is accounted for by variability in another. In a perfect correlation where $r = +1$ or -1, a variation in one of the variables is exactly matched by a corresponding variation in the other. This situation is rare in biology because many factors govern relationships between variables in organisms. Thus, r^2 indicates to what extent other factors are influencing x and y. In Example 14.3 the coefficient of determination is $0.8386^2 = 0.703$ or 70.3%. It follows that some 30% of the variation in mass is *not* accounted for by variation in otolith length. The mass of a fish is undoubtedly influenced by how much food there is in its stomach or whether it has just spawned, factors which are quite independent of otolith length. In Example 14.4, $r^2 = 0.9815^2 = 0.963$. Over 96% of the variation in fish length is accounted for by variation in otolith length. Therefore, the size range of otoliths in seal droppings or seabird regurgitates is a more useful indicator of the range of the *length* rather than of the *mass* of the fish in the diet.

14.8 Advice on measuring correlations

1. When observations of one or both variables are on an ordinal scale, use the Spearman Rank Correlation Coefficient. The number of units in a sample, that is, the number of paired observations should be between 7 and about 30. The ranking of over 30 observations is extremely tedious and is not commensurate with any marginal increase in accuracy.

   Where there are tied observations, proceed as explained in Section 3.6. The rank correlation coefficient becomes unreliable if more than about half the ranks are tied.

2. When observations are measured on interval scales the use of the product moment correlation coefficient should be considered. Sample units must be obtained randomly and the data should be *bivariate normal*, that is to say, both x and y observations should be approximately normally distributed. This can be checked as we suggest in Section 9.6. The broad outline of points in a scattergram of bivariate normal data is roughly circular or elliptical. The circle becomes drawn out into an ellipse as r increases in value. Do not attempt to measure correlations of *bimodal* data (see Section 5.4). If possible, separate the observations attributable to each mode, for example, male and female, before analysing correlations.

3. The relationship between two variables should be rectilinear, *not* curved. A scattergram will show if this is the case. Certain mathematical transformations will 'straighten up' curved relationships to allow r to be calculated. The logarithmic transformation for counts and the arcsine transformation for proportions are commonly used. We discuss these further in Section 15.12.

4. Do not conclude that because two variables are strongly and significantly correlated that one is necessarily the *cause* of the other. It is always possible that some additional, unidentified factor is the underlying source of variability in both variables.

5. Correlations measured in samples estimate correlations in the populations from which they are drawn. A correlation in a sample is not improved or strengthened by obtaining more sample units; however, larger samples may be required to confirm the statistical significance of weak correlations.

15 REGRESSION ANALYSIS

15.1 Introduction

In Section 4.6 we illustrated the relationship between the mass and the length of a sample of animals by means of a scattergram. In presenting a scattergram it is often helpful to draw a line through the cloud of points in such a way that the *average* relationship is depicted. The line is called the **line of best fit**. It has been added to the scattergram in Fig. 15.1. A problem arises as to how to fit the best line through a cloud of points. If the scatter is not too great, the line may reasonably be fitted 'by eye'. In most cases, however, it is necessary to replace such a subjective method by a more objective mathematical approach. The line so produced is called a **regression line**. The regression line may be described in terms of a mathematical equation which defines the relationship between x and y and we may use this equation to estimate or predict the value of one variable from a measurement of the other. We call this procedure (i.e. fitting the best line to a scattergram from an equation relating x and y) **regression analysis**. Before we attempt to describe this important statistical technique we must first consider a little basic geometry.

15.2 Gradients and triangles

The gradient of a hill slope is often expressed in such terms as 1 in 10. This means simply that for every 10 units of distance travelled in a horizontal plane an elevation of 1 unit in the vertical plane will result. The gradient, or slope, is symbolized by b and, in this case, is equal to $1 \div 10 = 0.1$. In the general case, the gradient is equal to an increment in y divided by a corresponding increment in x (Fig. 15.2).

Knowing the value of b we may use it to calculate the height gained from a given distance moved horizontally, thus:

Vertical height gained = gradient × horizontal distance travelled

In the general case, this may be expressed as:

$$y = bx$$

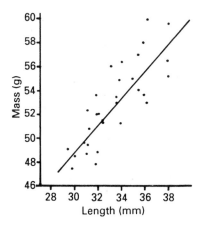

Fig. 15.1 A regression line.

If we wish to know the actual final height rather than just the height gained we must know the height of our starting point above some reference zero, say sea level. If this height is symbolized by a units, the final height will be:

Final height = height of starting point + (gradient × horizontal distance travelled)

or, in the general case (see Fig. 15.3),

$$y = a + bx$$

The equation, $y = a + bx$, is known as the equation of a straight line or the **rectilinear equation**. Regression analysis is concerned with solving the values of a and b in the equation from a set of bivariate sample data. We may then accurately fit a line to a scattergram and estimate the value of one variable from a measurement of the other. The quantities a and b are both **regression coefficients**. In common usage, however, the term 'regression coefficient' is taken to mean the slope of the regression line, b.

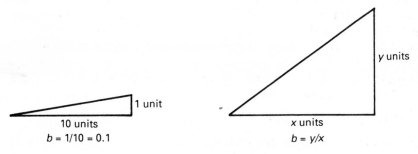

Fig. 15.2 Gradients of slopes: gradient = y units/x units.

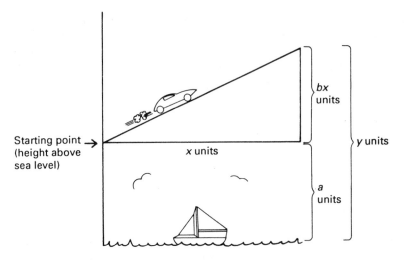

Fig. 15.3   Final height of car above sea level $= a + bx$ where b is the gradient.

15.3   Dependent and independent variables

To this point we have not indicated which of two variables should be placed on the *y* (vertical) axis and which on the *x* (horizontal) axis. There is a convention which gives us a guideline in this respect. In many pairs of variables it is possible to discern, unambiguously, that one of the variables is dependent on the other. For example, the number of warblers breeding in a county might depend on the length of hedgerow available. The notion that the converse might be true, that the length of hedgerow in some way depends on the number of warblers, is clearly ludicrous. In other examples we have considered, the diversity of grasshoppers (Example 14.1) in meadows appears to depend on the time elapsed since pesticides were used. Likewise, the number of stonefly nymphs in streams (Example 14.2) appears to be dependent (negatively) on the hardness of the water. The height gained by the car in Section 15.2 depends on the horizontal distance moved. The converse of these cannot be true. In these instances the variable which describes the number of warblers, the diversity of grasshoppers, the number of stoneflies and the height of the car is the **dependent variable**. In each case the other variable is the **independent variable**. In a scattergram or during regression analysis, we conventionally place the dependent variable on the *y*-axis and the independent variable on the *x*-axis.

We should note here that in identifying a dependent variable we are not admitting a *causal* relationship between the two. It is always possible that the two variables are independently related to a third, unidentified, variable.

Sometimes it is not possible to state which is the dependent and independent variable in a pair. Such is the case in the otolith length/fish length relationship of Example 14.3. If the regression line is to be used to estimate the value of one variable from a measurement of the other, then the variable which is used to estimate the other is placed on the x-axis; the variable *to be* estimated is placed on the y-axis. If, however, the regression line is calculated merely to describe the mathematical relationship between two variables, and no dependent variable can be identified, then the choice of axis is arbitrary.

5.4 A perfect rectilinear relationship

By example, we may relate the rectilinear equation derived in Section 15.2 to the idea of dependent and independent variables.

Example 15.1

An observer measures the length of a spring when different known masses are suspended from it. The following data are recorded:

Mass x (g)	Length of spring y (cm)
10	10
20	15
30	20
40	25
50	30

Clearly the length of the spring is dependent on the mass attached. We have no difficulty therefore in identifying mass as the x variable and length as the y variable. A scattergram drawn from these data is shown in Fig. 15.4a. We see that all points are in perfect alignment and that the regression line may be drawn through them without the need for a mathematical computation (Fig. 15.4b). When extrapolated downwards, the line cuts the y-axis at point a. This is the *intercept on the y-axis*. It represents the length of the spring in its unstretched state and is analogous to the *height above sea level* in the example in

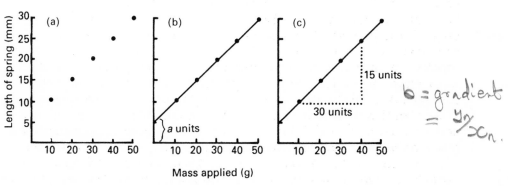

Fig. 15.4   A perfect linear relationship.

Section 15.2. We see from the scale on the *y*-axis that *a* has a value of 5 cm. Thus one of the quantities in the $y = a + bx$ equation has been determined.

To calculate the other, *b*, drop a vertical line from any point on the regression line and complete a right-angled triangle with a horizontal line as shown in Fig. 15.4c. The gradient, *b*, is the number of units on the *y*-scale divided by the number of units on the *x*-scale which correspond to the respective sides of the triangle. In Fig. 15.4c, $b = 15/30 = 0.5$.

We now have both quantities in the equation $y = a + bx$. Thus:

$$y = 5 + 0.5x$$

Using this equation we may now estimate the length of the spring when a mass of any size is attached. We can check this for a value we already know. An observation in the table above shows that a mass of 50 g produces a length of 30 cm. Does the equation bear this out?

$$y = 5 + (0.5 \times 50) = 5 + 25$$
$$y = 30 \text{ cm}$$

Clearly, we have obtained the correct result for the values of *a* and *b*. We may now confidently interpolate between the points and predict the length for, say, a mass of 25 g.

$$y = 5 + (0.5 \times 25) = 17.5 \text{ cm}$$

In this way we can build up a series of *y* values to produce a finely graded scale. If the scale were to be etched onto a stick aligned vertically beside the spring we would have a rudimentary spring balance.

Although it is safe to interpolate between points on the scattergram caution should be exercised in extrapolating far beyond the last point. If a 500 g mass is hung from the end of our spring it might well overload the capacity of the spring to return to its original length, and it is by no means certain that its stretched length is equal to that predicted by the equation.

As we have already noted, such perfect rectilinear relationships may be encountered in physics but seldom in biology, where a cloud of points rather than a nice straight line is more usual.

15.5   The line of least squares

Fitting a regression line to a scattergram involves placing it through the points so that the sum of the vertical distances (**deviations**) of all points from the line is minimized. Because some deviations are negative and some positive, it is more convenient to utilize the sum of the *squares* of the deviations, Σd^2. In this way awkward negative signs are eliminated. The method of fitting the line is therefore known as the **method of least squares**. Figure 15.5 shows the line of least squares fitted to four points in which the squares of the vertical deviations are minimized. Any alternative position of the line (up or down, or with different slope) will increase Σd^2. Although it is possible to reduce the values of

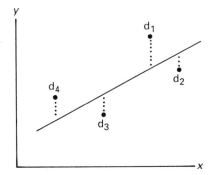

Fig. 15.5 Deviations from the regression line.

d_1^2 and d_4^2 by moving the line closer to these points, it is only at the expense of increasing $d_2^2 + d_3^2$. Since we are dealing with *squares*, the increase in $(d_2^2 + d_3^2)$ is not offset by the decrease in $(d_1^2 + d_4^2)$.

Fitting the line in this way, where Σd^2 for vertical deviations (deviations on the *y*-axis) is minimized, is called the **regression of y on x**. It allows for the estimation of *y* values from nominated values of *x*, and is usually referred to as **simple linear regression**.

5.6 Simple linear regression

Regrettably, the above title belies some rather strict conditions which apply to the application of least squares regression. The main conditions are:

(a) There is a linear relationship between a dependent *y* variable and an independent *x* variable which is implied to be *functional* or *causal*.
(b) The *x* variable is not a random variable but is under the control of the observer. Regression is therefore especially applicable to *experimental* situations in which, for example, the response of a *y* variable to predetermined quantities of time, temperature, mass of chemical, etc. is under investigation.
(c) For any single defined observation of an *x* variable there is a theoretical population of *y* values. The population is normally distributed and the variances of different populations of *y* values corresponding to different individual *x* values are similar (Fig. 15.6).

If these conditions are not met, then Model 2 regression (Section 15.15) should be considered.

Example 15.2

A biologist investigates the effect of applying different amounts of fertilizer on the yield of grass on reclaimed derelict land. Grass seed is sown uniformly over

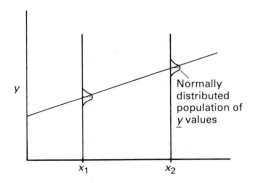

Fig. 15.6   Normally distributed populations of *y* values.

the area. Ten 1 m² plots are located randomly and a different mass of commercial fertilizer is applied evenly to each. Two months later the grass is carefully harvested from each plot, dried and weighed. The results of the experiment are tabulated below.

x variable: Mass of fertilizer (g/m²)	25	50	75	100	125	150	175	200	225	250
y variable: Yield of grass (g/m²)	84	80	90	154	148	169	206	244	212	248

(handwritten annotations above x variable row: 2100 4000 6750 15400 18500 25350 36050 48800 47700)

A scattergram of these data (Fig. 15.7) reveals a roughly rectilinear relationship; within the limits of the experiment the yield of grass clearly depends on the amount of fertilizer applied and a functional relationship is presumed. Observations on the *x*-axis are under the control of the observer and so simple linear regression is appropriate. Notice that the *x* observations are not normally distributed – they are evenly spread over a range of 25–250 g rather than being clustered around a mean value.

The information we require to calculate *a* and *b* is the same as that needed to calculate the correlation coefficient. In addition we must obtain the mean of the *x* observations ($\bar{x}$) and the mean of the *y* observations ($\bar{y}$). Using a calculator we determine:

$$\bar{x} = 137.5 \qquad \bar{y} = 163.5 \qquad n = 10$$
$$\Sigma x = 1375 \qquad \Sigma y = 1635 \qquad \Sigma xy = 266\,650$$
$$(\Sigma x)^2 = 1\,890\,625 \qquad (\Sigma y)^2 = 2\,673\,225$$
$$\Sigma x^2 = 240\,625 \qquad \Sigma y^2 = 304\,157$$

(handwritten: Steps to calculate b)

The computation of *a* and *b* for the regression of *y* on *x* requires two steps. First, the calculation of *b*, and then, using the value of *b* so derived, the calculation of *a*. The formula for the calculation of *b* is:

$$b = \frac{n\Sigma xy - \Sigma x \Sigma y}{n\Sigma x^2 - (\Sigma x)^2}$$

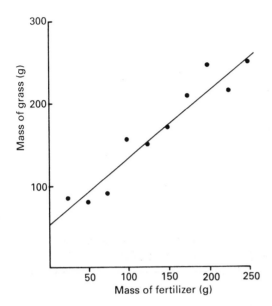

Fig. 15.7 Regression of mass of grass yield on mass of fertilizer applied.

Substituting in the formula:

$$b = \frac{(10 \times 266\,650) - (1375 \times 1635)}{(10 \times 24\,0625) - 1\,890\,625} = \frac{418\,375}{515\,625}$$

$b = 0.8114$ Calculate a)

To calculate *a* we use the rectilinear equation, $y = a + bx$ or, rearranging, $a = y - bx$. We have derived a value for *b*, but which of the 10 possible values of *x* and *y* should we use to solve the equation for *a*? Since all the points constitute a *scatter* it is most unlikely that any one single pair will be correct. The answer is to use the mean of *x* and the mean of *y* thus:

$$a = \bar{y} - b\bar{x}$$

We now have all the terms needed to solve for *a*:

$a = 163.5 - (0.8114 \times 137.5)$
$a = 163.5 - 111.5675$
$a = 51.933$

The regression equation may now be written out in full: fit into equation

$y = 51.933 + 0.8114x$

We can use the equation to predict the yield of grass from any given application of fertilizer. Thus, for an application of 115 g we predict:

Yield of grass (g) $= 51.93 + (0.8114 \times 115)$
$\qquad\qquad\qquad = 145.2\,\text{g}$

It is safe only to make predictions within the range of the initial x observations. The regression line undoubtedly curves off when the amount of fertilizer ceases to be the limiting growth factor. Indeed, the yield of grass may actually fall when the amount of fertilizer applied is so large as to become toxic.

The distribution of observations on the x-axis is not normal because they span a range and are not clustered around an average value. We are therefore unable to test the *significance* of this relationship by means of the product moment correlation coefficient. The procedure for testing the significance of a regression line is described in Section 15.10.

15.7 Fitting the regression line to the scattergram

To fit a line to a scattergram we need to know the positions of two points on the line. The further apart the points are, the more accurately can the line be drawn. The points are calculated from the regression equation, $y=a+bx$. Two separate values of x are selected to derive corresponding y values from the equation. These become the coordinates for the two points.

Select a value of x that is close to the y-axis. In Example 15.2 this could be 25 g. Calculate the corresponding y value from the equation:

$$y=51.93+(0.8114 \times 25)=72.2$$

Mark a point at coordinates $x=25$; $y=72.2$. If the x axis is scaled to start at zero, then the y-coordinate at $x=0$ will of course equal a, the intercept. The point can be marked here on the y-axis. Select a second value of x, far from the y-axis, say 250 g. As before, calculate the corresponding y value from the equation:

$$y=51.93+(0.8114 \times 250)$$
$$y=254.8$$

Mark a second point on the scattergram at coordinates $x=250$; $y=254.8$. Join the two points with a straight line but do not exceed the cloud of points by very much at each end.

15.8 The error of a regression line

From the regression equation of Example 15.2 we estimate that an application of 115 g of fertilizer results in a yield of 145.2 g of grass grown under the standard conditions. Can we say how good an estimate this is? We recognize, first, that the regression line is derived from *sample* data and is therefore subject to the same sort of sampling error that occurs in the estimation of a mean or any other population parameter. Suppose that we apply 115 g fertilizer to 1000 plots of 1 m²; that is to say, a *population* of plots. The yields of grass from the plots constitute a normal distribution of y values, all corresponding to the x

value of 115 g. The mean of this distribution is the true population estimate and is the y coordinate through which the regression line passes without error. The standard deviation of the distribution is the **standard error (S.E.) of the estimate**. How the standard error is calculated is described below. Applying the same argument as in Section 11.4, we are 95% confident that the *population* value falls within $\pm (t \times S.E.)$ of an estimate based on a *sample* regression. Thus, in Example 15.2 the y-coordinate of the true regression line at $x = 115$ g is within $145.2 \pm (t \times S.E.)$. We obtain the value of t at the appropriate degree of freedom from Appendix 2. Then a confidence interval can be placed vertically above and below the regression line at $x = 115$. By repeating the procedure for several values of x, a **confidence zone** is established on either side of the regression line. The upper and lower boundaries of the zone are not parallel to the regression line but fan out towards each end.

Before the standard error is obtained, a quantity called the **residual variance**, s_r^2 is first calculated using the formula:

$$s_r^2 = \frac{1}{n-2} \times \left(\text{sum of squares of } y - \frac{(\text{sum of products})^2}{\text{sum of squares of } x} \right)$$

We solve, in turn, each item within the large brackets.

(a) The sum of products (SP) of x and y is explained in Section 14.5 and is written as $\Sigma(x - \bar{x})(y - \bar{y})$. A more direct way of obtaining it is by using the alternative formula:

$$SP_{x,y} = \Sigma xy - \frac{\Sigma x \Sigma y}{n} \quad \text{sum of products}$$

Inserting the terms we derived in Example 15.2

$$SP_{x,y} = 266\,650 - \frac{1375 \times 1635}{10} = 41\,837.5$$

Therefore, $(SP_{x,y})^2 = 41\,837.5^2 = 1\,750\,376\,406$

(b) The sum of squares (SS) of y is given by (Section 6.7):

$$SS_y = \Sigma y^2 - \frac{(\Sigma y)^2}{n} = 304\,157 - \frac{2\,673\,225}{10} = 36\,834.5$$

(c) In the same way, the sum of squares of x is given by (Section 6.7):

$$SS_x = \Sigma x^2 - \frac{(\Sigma x)^2}{n} = 240\,625 - \frac{1\,890\,625}{10} = 51\,562.5$$

(d) Substitute these three numbers in the formula for residual variance:

$$s_r^2 = \frac{1}{8} \times \left(36\,834.5 - \frac{1\,750\,376\,406}{51\,562.5} \right) = \frac{1}{8} \times 2887.806 = 360.98$$

To calculate the standard error of a point on the regression line corresponding to an estimate of y (designated y') from a stated value of x (designated x'), the formula is:

$$S.E. = \pm \sqrt{s_r^2 \times \left[\frac{1}{n} + \frac{(x' - \bar{x})^2}{SS_x}\right]}$$

Using our regression equation from Example 15.2, a value of x' of 115 g of fertilizer gives an estimate, y', of 145.2 g yield of grass. Substituting these values in the equation for the standard error (with SS_x derived in (c) above):

$$S.E. = \pm \sqrt{360.98 \times \left[\frac{1}{10} + \frac{(115 - 137.5)^2}{51\,562.5}\right]}$$

$$= \pm \sqrt{360.98 \times 0.10\,982}$$
$$= \pm 6.296$$

To convert the standard error to a 95% confidence interval we multiply by the appropriate value of t (see Section 11.4). Consulting Appendix 2 at $P = 0.05$ the tabulated value of t at $(n-2) = 8$ df is 2.306. The 95% confidence limits (C.L.) about y' are therefore $y' \pm t \times S.E.$

95% C.L. $= 145.2 \pm 2.306 \times 6.296$
$= 145.2 \pm 14.519$
$= 159.72$ (upper limit) and 130.681 (lower limit)

We are therefore 95% confident that if we were to replicate the experiment an infinite number of times (thereby producing a *population* regression line), the point on the line corresponding to $x = 115$ will fall between $y = 159.72$ and $y = 130.681$.

Computation of the 95% confidence limits for several values of x over the range of the x-scale will enable the 95% confidence zone to be plotted. Because of the fanning out effect, it is not possible to define the zone from only two values of x, as it was when drawing the regression line itself (Section 15.7).

The 95% confidence zone of our regression line is shown as the *inner* zone in Fig. 15.8.

15.9 Confidence limits of an individual estimate

The 95% confidence zone described in the previous section is the confidence zone for the line as a whole. Predicted values of y for *individual* values of x are subject to an additional source of error, namely scatter about the regression line. We must therefore establish a second confidence zone whose limits are further out from the limits of the confidence zone of the regression line.

If the 95% confidence limits for the regression line have been calculated, those for individual estimates are determined by adding 1 in the square root term in the equation used for calculating the limits of the regression line. The

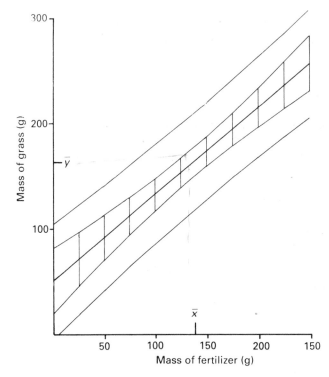

Fig. 15.8 Confidence zones of a regression line.

zone may be determined by similarly working out the confidence limits for several values of x over the range of the x-scale. Thus:

$$95\% \text{ C.L.} = y' \pm t \times \sqrt{s_r^2 \times \left[1 + \left(\frac{1}{n}\right) + \frac{(x' - \bar{x})^2}{SSx}\right]}$$

Using the same values in the previous section, the 95% confidence limits about the predicted yield from a single application of 115 g of fertilizer is:

$$
\begin{aligned}
95\% \text{ C.L.} &= 145.2 \pm t \times \sqrt{360.98 \times (1 + 0.10\,982)} \\
&= 145.2 \pm (2.306 \times 20.016) \\
&= 145.2 \pm 46.156 \\
&= 191.356 \text{ g (upper limit) and } 99.04 \text{ g (lower limit)}
\end{aligned}
$$

What this means is that if we were to make a *single* application of 115 g of fertilizer to a 1 m² plot under the standard conditions we are 95% confident that the yield of grass will be between 191.36 g and 99.04 g. The limits are shown in Fig. 15.8 and represent a considerable extension to the breadth of the confidence zone of the regression line.

15.10   The significance of the regression line

[handwritten: Probability that there is a true linear relationship between x and y ...]

If the scatter of points about the regression line is considerable, and the value of the regression coefficient, b, is low, it may be necessary to test the significance of the regression. This will tell us the probability that there is a true linear relationship between the x and y variables in the parent population.

If there is no relationship, the value of b is zero. That is, the slope of the regression line is parallel to the x-axis. The test for significance is therefore a test for a significant departure of the value of b from zero. The test depends on the computation of a t value (at $n-2$ df) from the values of b and the residual variance, s_r^2, whose derivation was described in Section 15.8.

The steps involved in the calculation of t are as follows. The three quantities needed are s_r^2 (360.98 from Section 15.8); the sum of squares of x (51 562.5 from Section 15.8); and the value of the slope, b (0.8114 from Section 15.6).

Step 1
Work out the standard error (S.E.) of b from the formula:

$$\text{S.E.}_{b} = \sqrt{\frac{\text{residual variance}}{\text{sum of squares of } x}}$$

$$= \sqrt{\frac{360.98}{51\,562.5}}$$

$$= 0.08367$$

Step 2
Estimate t from:

$$t = \frac{b}{\text{S.E.}_{b}} = \frac{0.8114}{0.08367} = 9.698 \text{ with } (n-2) = 8 \text{ df}$$

Step 3
Compare the calculated value with the tabulated value at 8 df in Appendix 2. The calculated value greatly exceeds the value of 3.355 at $P = 0.01$. The linear relationship between the y and x variables is highly significant.

Alternatively, significance may be tested by analysis of variance (Section 17.11).

15.11   The difference between two regression lines

Sometimes we need to know if there is a significant difference between the slopes of two regression lines. Suppose that the biologist in Example 15.2

conducts a second experiment alongside the first, using a different supply of fertilizer. We assume that 10 quadrats are again treated with the same quantities of fertilizer; the slope, b of the new regression line is found to be 0.5221 with a standard error of 0.08831.

Summarizing the sample data:

$b_1 = 0.8114$ $S.E._{.b1} = 0.08367$
$b_2 = 0.5221$ $S.E._{.b2} = 0.08831$

The value of the second slope is less than that of the first, but is the difference simply due to the *scatter* within the samples?

$H_0 =$ the difference between the slopes is attributable to sampling error
$H_1 =$ the difference is too large to be attributed to sampling error (*two-tailed* test).

The steps in the procedure are as follows:

Step 1
Obtain the difference between b_1 and b_2

$$(b_1 - b_2) = (0.8114 - 0.5221) = 0.2893$$

Step 2
Obtain the *standard error of the difference*

$$S.E._{.(b1-b2)} = \sqrt{S.E._{.b1}^2 + S.E._{.b2}^2} = \sqrt{0.08367^2 + 0.08831^2}$$
$$= 0.1216$$

Step 3
Estimate t from:

$$t = \frac{(b_1 - b_2)}{S.E._{.(b1-b2)}} = \frac{0.2893}{0.1216} = 2.379$$

Step 4
Work out the degrees of freedom

$$df = (n_1 - 2) + (n_2 - 2) = (8 + 8) = 16$$

Step 5
Consult Appendix 2 at 16 df.

The calculated value of 2.379 is larger than the tabulated value of 2.120 at $P = 0.05$ but smaller than 2.921 at $P = 0.01$. The difference is statistically significant, but not highly significant.

In a two-tailed test, any negative sign associated with the estimated t value is ignored. If there is an underlying reason to suppose that the value of one slope will be less than another, then the test is *one-tailed*. Nominate the slope which is predicted to be the larger b_1 and use the one-tailed columns in Appendix 2. We suggest you always use the two-tailed version unless there is good reason for doing otherwise.

15.12   Dealing with curved relationships

We noted in Section 15.6 that one of the conditions applicable to simple linear regression is that there is a *linear* relationship between the *y* and *x* variables. Many strong relationships in biology are not linear but exhibit *curved* lines of best fit. Growth and mortality is an example of this. The usefulness of regression is greatly increased when curved relationships are 'straightened up' by *transformation*. It then becomes possible to undertake regression analysis and to estimate a value of one variable from that of another.

Example 15.3

In a study of amphibian mortality, a mass of developing frog spawn is released into a suitable environment. At two-week intervals, predetermined by the observer, the numbers of surviving tadpoles are counted. The relationship between numbers surviving and time is shown in Fig. 15.9(a); it is distinctly curved. In this case the curve is straightened up by transforming the counts of tadpoles to logarithms, that is, replacing each value of *y* by log *y*. In Fig. 15.9(b) we see that this mathematical device has straightened up the curve.

The transformation appears to be satisfactory. We may therefore proceed with regression analysis provided that all *y*-values are replaced by their logarithms. The data are presented in the table below. We designate the numbers of tadpoles *y* and their logarithms *y'* in order to distinguish between them.

Week number (x)	Number of tadpoles surviving (y)	Log (number of tadpoles surviving) (y')
2	541	2.7332
4	116	2.0645
6	58	1.7634
8	27	1.4314
10	6	0.7782
12	3	0.4771

Using a calculator to process and summarize the data:

$\Sigma x = 42$ $\Sigma y' = 9.248$ $n = 6$

$(\Sigma x)^2 = 1764$ $(\Sigma y')^2 = 85.526$ $\Sigma x y' = 49.2632$

$\Sigma x^2 = 364$ $\Sigma y'^2 = 17.724$

$\bar{x} = 7.0$ $\bar{y}' = 1.5413$

Substituting in the formula for *b* (Section 15.6):

$$b = \frac{[(6 \times 49.2632) - (42 \times 9.248)]}{(6 \times 364) - 1764} = \frac{-92.837}{420}$$

$b = -0.2210$

The negative value of b indicates the declining slope of the graph. Solving for a as before, but remembering that this time a is measured on a logarithmic scale, we designate it a':

$a' = \bar{y}' - b\bar{x} = 1.5413 - (-0.2210 \times 7.0)$
$a' = 3.088$

We may now substitute the coefficients into the regression equation:

$y' = a' + bx$

That is,

$y' = 3.088 + (-0.2210 \times x)$

Using this equation to estimate the number of tadpoles surviving after 3 weeks $(x = 3)$:

$y' = 3.088 + (-0.2210 \times 3) = 2.425$

To back-transform y' to y, take the antilog:

antilog $2.425 = 266$ tadpoles

An alternative way of writing the regression equation is:

$\log y = \log a + bx$

or, in exponential form:

$y = a \times 10^{bx}$ where a is antilog $a' = 1224.6$

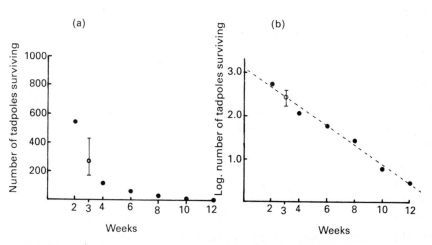

Fig. 15.9 Graph of survival of tadpoles. (a) Axes untransformed; (b) y-axis transformed logarithmically. Open circles with error bars are estimates and 95% confidence intervals at $x = 3$. The dashed line in (b) is the regression line $y' = 3.088 - (0.221x)$. See text for details.

Thus, the estimated number of tadpoles surviving after 3 weeks is:

$$y = 1224.6 \times 10^{(-0.2210 \times 3)}$$
$$= 1224.6 \times 0.2173 = 266 \text{ tadpoles}$$

The back-transformed intercept, a, is of course the estimated number of viable ova in the liberated spawn in week zero.

Example 15.4

Establish the 95% confidence limits to the regression line at $x = 3$ weeks. Proceed as described in Section 15.8, keeping the y observations in logarithms. Confirm that the residual variance s_r^2 is 0.0124, the standard error of b is 0.0133 and the confidence interval at $x = 3$ is 2.43 ± 0.20 ($t_{4df} = 2.776$). [Your answer may differ on the last decimal place or two due to rounding errors during computation.]

When back-transforming the confidence interval, remember that a number which is added and subtracted as a logarithm is multiplied and divided when converted to an ordinary number. Thus:

antilog $[2.43 \pm 0.20]$ is $269.153 \overset{\times}{\div} 1.585$, i.e. 426.6 (upper limit) and 169.8 (lower limit).

The interval is not symmetrically placed above and below the estimate. The estimate and the confidence interval are shown in Fig. 15.9.

15.13   Transformation of both axes

Logarithmic transformation of one or both of the axes of a curvilinear relationship will usually result in an adequate straightening. But how do we decide whether one of the two axes, or both, should be transformed? The answer may be: by trial and error. If a scattergram of untransformed bivariate data is curved, prepare scattergrams with first one axis transformed, then the other, then both, until the best result is obtained. The best outcome can usually be assessed by visual inspection of the scattergrams. If the scatter is considerable and there is doubt, then a more objective method is to derive the **coefficient of determination** r^2 (see Section 14.7) for each alternative transformation. The one which gives the largest value of r^2 is the one to use for regression.

Example 15.5

Island biogeographic theory tells us that the number of species on islands increases with the size of the island, but not in a rectilinear fashion. Table 15.1 shows the number of higher plant species found on eight islands in Shetland. Area is the x variable and the number of species the y variable. The coefficient of determination for each of the possible transformations is:

(i) y on $x: r^2 = 0.698$
(ii) $\log y$ on $x: r^2 = 0.458$
(iii) y on $\log x: r^2 = 0.912$
(iv) $\log y$ on $\log x: r^2 = 0.962$

The *double logarithmic* transformation gives the largest value of r^2. Table 15.1 includes the natural logarithm of each observation ($\ln x$ and $\ln y$).

Table 15.1   Number of higher plants on islands of different size

Island	Area x (ha)	ln area x'	No. species y	ln (No. species y')
Whalsay	1850	7.523	164	5.10
Hascosay	269	5.59	122	4.80
Samphrey	72	4.28	67	4.20
Uynarey	21	3.04	70	4.25
Orfasay	9.5	2.25	43	3.76
Gruney	7.0	1.95	37	3.61
Kay Holm	2.0	0.693	23	3.14
Tinga Skerry	0.375	−0.981	18	2.89

Proceeding with linear regression on the transformed observations and designating $\ln x$ as x' and $\ln y$ as y', the essential data are summarized:

$\Sigma x' = 24.345$ $\Sigma y' = 31.75$ $n = 8$
$(\Sigma x')^2 = 592.68$ $(\Sigma y')^2 = 1008.1$ $\Sigma x'y' = 110.94$
$\Sigma x'^2 = 125.71$ $\Sigma y'^2 = 130.13$
$\bar{x}' = 3.043$ $\bar{y}' = 3.97$

Solving for b (Section 15.6):

$$b = \frac{(8 \times 110.94) - (24.345 \times 31.75)}{(8 \times 125.71) - 592.68} = \frac{114.57}{413}$$

$b = 0.2774$

Solving for a':

$a' = (y' - b\bar{x}') = 3.97 - (0.2774 \times 3.043)$
$a' = 3.126$

Therefore,

$y' = 3.126 \times 0.2774x'$

Thus,

$\ln(\text{number of species}) = 3.126 + [0.2774 \times \ln(\text{area of island})]$

To estimate the number of species S on an island of area 100 ha:

$\ln(S) = 3.126 + (0.2774 \times \ln 100) = 4.403$

The estimated number of species is a number whose natural logarithm is 4.403, namely 81.7 (i.e. antilog 4.403). Since species can only occur in whole numbers, this is rounded to 82.

An equation in the form $\ln y' = \ln a' + b \ln x'$ may be back-transformed by taking antilogs on each side of the equation to derive:

$y = a \times x^b$, where a is antilog $a' = 22.78$

Thus the number of species expected on an island of 100 ha is:

$$S = 22.78 \times 100^{0.2774} = 22.78 \times 3.588$$
$$= 81.7$$

Confidence limits to estimates can be worked out as described in Sections 15.8 and 15.9 on the logarithmically transformed observations. Do not forget that confidence limits back-transformed from logarithms are multiplied and divided as explained in Example 15.4.

15.14   Regression through the origin

In some situations we might expect that as values on the x-axis decline towards zero, so do values on the y-axis. When this is the case, the regression line passes through zero on both axes, i.e. through the *origin*. Then, there is no intercept on the y-axis, and the value of a is accordingly zero. Since the origin is a point on the regression line that is without error, the regression equation is modified to 'force' the line through this point. The equation is:

$y = bx$

and b is solved from the simpler equation:

$$b = \frac{\Sigma xy}{\Sigma x^2}$$

Example 15.6

Thrips (Thysanoptera) are tiny insects which feed on pollen grains; their populations can sometimes grow to pest proportions. In an experiment to discover feeding rates, thrips are fed pollen grains of a range of sizes. The time taken to consume grains of known size is recorded. Since it is expected that a pollen grain of zero size will be consumed in zero time, regression through the origin is called for. Data are tabulated in Table 15.2.

$$b = \frac{2320}{2621} = 0.885$$

$y = 0.885x$

The estimated time for a thrips to consume a pollen grain of volume $25 \times 10^3 \ \mu m^3$ is $y = 0.885 \times 25 = 22.1$ s.

Table 15.2 Feeding times of thrips

Volume of grain, x ($\mu m^3 \times 10^3$)	x^2	Mean consumption time per grain, y (s)	xy
40	1600	35	1400
20	400	18	360
16	256	15	240
15	225	12	180
10	100	11	110
6	36	4	24
2	4	3	6
	$\Sigma x^2 = 2621$		$\Sigma xy = 2320$

To draw the regression line, estimate a value of y from any single value of x, using $y = 0.885\,x$. The line joining the point at these coordinates with the origin (zero point on both axes) is the regression line.

15 An alternative line of best fit

In Section 15.6 we outline some of the conditions that have to be met if regression by least squares is to be correctly applied. One of the important conditions is that the x variable is not a *random variable* but is fixed or controlled by the observer. In many instances where biologists wish to place a line of best fit through points on a scattergram, this condition is not met. In Examples 14.3 and 14.4 we analysed bivariate data which reveal a strong correlation between the length of a fish and the length of an otolith dissected from it. Here, both variables – fish length and otolith length – are random variables. Neither is under the control of the observer. It is, of course, mathematically possible to follow the procedure followed in Section 15.6 to calculate the regression coefficients a and b. However, estimates of these coefficients are statistically biased. In circumstances like this, an alternative form of regression known as **Model 2** is applied. There are a number of versions of Model 2 regression (see Sokal and Rohlf, 1981, Section 14:13). By far the simplest to apply is that known as **reduced major axis regression**.

The slope of the regression line (symbolized b' to distinguish it from least squares regression) is given by:

$$b' = \pm \frac{\text{standard deviation of } y \text{ observations}}{\text{standard deviation of } x \text{ observations}}$$

The notation '$\pm$' is placed in front of the ratio because the formula does not give the *sign* of the slope. Standard deviations it should be noted, are always positive. The sign of the slope, $+$ or $-$, is decided by inspection of the scattergram. If the points are too scattered and the direction of the slope not obvious, then the sign of the *sum of products* (Section 15.8) is the sign of the slope. A word of caution however. If the direction of a slope is not obvious in a scattergram of the data, then regression analysis will be of no practical value for making estimations.

The intercept, a' is estimated in the usual way:

$$a' = (\bar{y}' - b'\bar{x})$$

Two further points should be held in mind when applying Model 2 regression.

(i) There is no requirement that sampling units should be obtained randomly. Indeed, it is preferable to select items which span the available range for measurement. The biologist in Examples 14.3 and 14.4 would be advised to sort through the fish box to choose some small, medium-sized and large fish. The technique may nevertheless be applied to randomly sampled observations.

(ii) Because both variables are random variables, there is no clear dependent variable. The x and y axes are therefore assigned arbitrarily.

Example 15.17

A biologist obtains eight cod from a fisherman. The fish are measured (y variable) and dissected to remove an otolith which is also measured (x variable). The observations are:

Otolith length, x (mm)	28.0	18.1	14.8	13.1	12.6	11.0	9.2	7.7
Fish length y, (mm)	737	416	312	280	249	215	156	117

Using a scientific calculator we determine:

Mean of x data, $\bar{x} = 14.3125$; standard deviation of x data $s_x = 6.405\,006$
Mean of y data, $\bar{y} = 310.25$; standard deviation of y data $s_y = 195.7329$

$$b' = \pm\frac{s_y}{s_x} = \frac{195.7329}{6.405\,006} = 30.559$$

From an inspection of the data it is obvious that as the length of fish increases so too does the length of otolith. The sign of the slope is therefore positive.
Solving for a':

$$a' = (\bar{y} - b'\bar{x}) = 310.25 - (30.559 \times 14.3125)$$
$$= -127.126$$

(The negative value of a' means simply that the regression line intercept on the y-axis is below the intersection of the x-axis.)

The equation is: $y = -127.126 + 30.559x$

This equation defines the mutual slope of the two random variables (fish length and otolith length). It should be used to draw the line of 'best fit' through a scattergram of such data.

The *significance* of the line may *not* be tested by the method described in Section 15.10. A *not significant* result indicates that b' does not depart from zero, implying that one of the standard deviations is zero. That is not possible. Significance is tested by analysis of variance (Section 17.11).

In practice, biologists may wish to make estimations (or predictions) from such data. They might, for example, wish to estimate the length of a fish eaten by a seabird or seal from the length of an otolith recovered in a regurgitate or a dropping. When an equation is to be used for making estimations of this sort (rather than for merely fitting a line to a scattergram) statisticians consider that simple linear regression (Section 15.6) with its attendant confidence zones (Sections 15.8 and 15.9) is more reliable. Then, the variable *to be* estimated (fish length in this example) is *always* assigned as the y variable.

.16 Advice on using regression analysis

1. A regression equation can be derived for any set of bivariate data simply by substituting them in the formulae. Because we *can* undertake a particular mathematical treatment, however, it does not mean that we *should*. The use of regression analysis should be restricted to cases where it is necessary to place a *best line* through a cloud of points and, in particular, where the estimation of one variable from a measurement of the other is required. It is a common fallacy to imagine that because an equation has been solved, the quality of the analysis is somehow enhanced. If all that is required is a measure of the strength of the relationship between two variables the use of the correlation coefficient may be more appropriate.
2. Before proceeding with the mathematics of regression, always draw a scattergram of the data to see what it *looks* like. From this you can decide if two important conditions are met: (i) Are the points roughly linear, not curved? and (ii) is the scatter of points around the line reasonably even over the whole length of the line?
3. If the scattergram suggests a *curved* relationship then transformation of one or other or both axes will usually straighten up the line. We have given examples of logarithmic transformations, but arcsine transformation for proportions or square-root transformation may prove effective. Many readers will be aware of the classical regression known as the Lineweaver–Burke plot in which the curved relationship between velocity of an enzyme-catalysed reaction and substrate concentration is straightened up by a *reciprocal* transformation of both axes; that is, x is replaced by $1/x$ and y by $1/y$.

4. The correlation coefficient *r* has no meaning in regression unless sample units have been obtained randomly and observations are normally distributed on both axes; that is to say, they are *bivariate normal*. However, the *coefficient of determination* r^2 is a useful number. It tells us the proportion of the variability in the *y* observations that can be accounted for by variability in the *x* observations (see Section 14.7). It is *not* an index of significance.

5. The *significance* of a regression line is determined by testing for a significant departure of *b*, the slope, from zero. Alternatively, it may be determined by analysis of variance (Section 17.11). Analysis of variance is the only means of testing significance in *Model 2* regression.

6. In reporting the results of your regression analysis remember to say which method you use.

6 COMPARING AVERAGES

6.1 Introduction

Biologists often wish to compare the average value of some variable in two samples. The problem typically resolves into two steps:

(i) is the observed difference between the average of two samples *significant*, or is it due to the chance sampling error that we described in Section 12.3?
(ii) if a difference between the average of two samples is indeed *significant*, what is the extent of the difference? *SE diff.*

We have already dealt with the second step in Sections 11.6 and 11.7. In this chapter we outline methods which test the significance of a difference between the average of two samples. In each method we describe, the Null Hypothesis is similar:

H_0 (two-tailed test): samples are drawn from populations with identical averages and any observed difference between the samples is due to sampling error.

The alternative hypotheses are:

H_1 (two-tailed test): samples are drawn from populations with different averages; an observed difference between samples cannot be accounted for by sampling error.

Or,

H_1 (one-tailed test): a nominated sample is drawn from a population with a larger average than that from which the other is drawn.

Tests for difference between sample averages may be non-parametric or parametric. Non-parametric tests convert observations to ranks and compare sample distributions, essentially their *medians*. Parametric tests use actual observations and compare *means*. We describe non-parametric methods first and then proceed to parametric methods.

6.2 Matched and unmatched observations

When analysing bivariate data such as correlations, a single sample unit gives a pair of observations representing two different variables. The observations comprising a pair are uniquely linked and are said to be *matched*. It is also possible to obtain a pair of observations of a *single* variable which are matched. For example, the masses of 10 birds at one site and the masses of another 10 at another site are unmatched. However, the masses of 10 birds on one day and the masses of the same 10 birds, identified by their ring numbers a week later are matched because the mass from one day can be paired off with that from the other. It is possible to conduct a more sensitive analysis if the observations are matched.

16.3   The Mann–Whitney *U*-test for unmatched samples

The Mann–Whitney *U*-test is a non-parametric technique for comparing the medians of two unmatched samples. It may be used with as few as four observations in each sample. Because the values of observations are converted to their *ranks*, the test may be applied to variables measured on ordinal or interval scales. Moreover, because the test is distribution-free, it is suitable for data which are not normally distributed, for example counts of things, proportions, or diversity indices. Sample sizes may be unequal. The task is to compute a test statistic *U* which is compared with tabulated critical values (Appendix 6).

Example 16.1

A biologist wishes to compare the average number of beetles captured in a sample of eight pitfall traps set in one woodland type, with that in a sample of seven traps set in another. Individual counts are listed below in ascending order for convenience.

Sample 1:	8	12	15	21	25	44	44	60
Sample 2:	2	4	5	9	12	17	19	

The median in Sample 1 is 23, considerably larger than the median of 9 in Sample 2 (see Section 5.3). However, there is considerable overlap between the observations in the samples. A test is required to decide if the difference is statistically significant. The Mann–Whitney *U*-test is appropriate. The procedure for using the test is as follows:

1. List all observations in both samples in ascending order, assigning them ranks. Where there are tied ranks, the average rank is assigned, as explained in Section 3.6. Distinguish between the samples by underlining one of them. We have underlined observations in Sample 1.

Observation:	2	4	5	8	9	12	12	15	17	19	21	25	44	44	60
Rank:	1	2	3	4	5	$6\frac{1}{2}$	$6\frac{1}{2}$	8	9	10	11	12	$13\frac{1}{2}$	$13\frac{1}{2}$	15

2. Sum the ranks of each sample; that is, sum the ranks of the underlined and non-underlined separately. Let $R_1 = $ sum of the ranks of Sample 1 and $R_2 = $ the sum of the ranks of Sample 2.

$$R_1 = 4 + 6\tfrac{1}{2} + 8 + 11 + 12 + 13\tfrac{1}{2} + 13\tfrac{1}{2} + 15 = 83.5$$
$$R_2 = 1 + 2 + 3 + 5 + 6\tfrac{1}{2} + 9 + 10 = 36.5$$

3. Calculate the test statistics U_1 and U_2 from

$$U_1 = n_1 n_2 + \frac{n_2(n_2 + 1)}{2} - R_2$$

$$= 56 + 28 - 36.5 = 47.5$$

$$U_2 = n_1 n_2 + \frac{n_1(n_1 + 1)}{2} - R_1$$

$$= 56 + 36 - 83.5 = 8.5$$

4. Check at this point that:

$$U_1 + U_2 = n_1 n_2$$
$$47.5 + 8.5 = 56 = 8 \times 7$$

This being the case, proceed to Step 5.

5. Select the *smaller* of the two U values (i.e. $U_2 = 8.5$ in this example) and compare it with the value in the table (Appendix 6) for the appropriate values of n_1 and n_2 (8 and 7 in this example). From the table, the critical value (at $n_1 = 8$ and $n_2 = 7$) is 10. Our smaller value of U is *less* than the critical value. The Null Hypothesis is therefore rejected. There is a statistically significant difference between the medians ($U = 8.5$; $P < 0.05$, Mann–Whitney U-test).

5.4 Advice on using the Mann–Whitney U-test

1. The Mann–Whitney U-test may be applied to interval data (measurements), to counts of things, derived variables (proportions and indices) and to ordinal data (abundance scales, etc.). It may be used with as few sampling units in each sample as 2 and 8, 5 and 3, or 4 and 4. However, with samples as small as these, there must be no overlap of observations between the two samples for H_0 to be rejected. Since this can be determined by inspection of the data it is hardly worth proceeding with the test.
2. Note that, unlike some test statistics, the calculated value of U has to be *smaller* than the tabulated critical value in order to reject H_0.
3. The test is a test for difference in *medians*. It is a common error to record a statement like 'the Mann–Whitney U-test showed there is a significant difference in means'. There is, however, no need to calculate the medians of each sample to do the test.
4. Although there is no requirement for the observations in the samples to be

normally distributed, the test does assume that the two distributions are similar. It is not permissible therefore to compare the median of a positively skewed distribution with that of a negatively skewed one. Since it is usually impossible to identify a frequency distribution in small samples the point is largely academic. Nevertheless, if it is known from other studies that the two samples have been drawn from populations which have fundamentally different frequency distributions then the test should not be used.

16.5   More than two samples – the Kruskal–Wallis test

At the end of Section 13.4 we identified a need for a test which compares the averages of several samples. Suppose that a biologist records the number of orchids counted in five randomly placed quadrats in each of four fields, A, B, C, D and then wishes to know if there are differences between them. It is possible to compare their medians using the Mann–Whitney U-test, but the test would have to be repeated six times to compare A with B, A with C and so on for all combinations. Apart from being extremely tedious, there is an important statistical reason for avoiding *multiple comparisons* of this type. We explain the reason fully in Section 17.1. The Kruskal–Wallis test is a simple non-parametric test to compare the medians of three or more samples. Observations may be interval scale measurements, counts of things, derived variables or ordinal ranks. If there are only three samples then there must be at least five observations in each sample. Samples do not have to be of equal size. The method of performing the test is explained in the following example.

Example 16.2

A biologist counts the number of orchids in five random quadrats in four meadows A, B, C, D. Are there differences between the average count in each field? The steps in performing the Kruskal–Wallis test are as follows and relate to Table 16.1.

1. Tabulate the observations in columns for each sample A, B, C, and D. Assign to each observation its *rank within the table as a whole*. If there are tied ranks assign to each its average rank as described in Section 3.6. Place each rank in brackets beside its observation.
2. Write the number of observations n in each sample (five in each in this example) in a line underneath its respective column. Add these up to obtain N, the *total* number of observations (20 in this case).
3. Sum the ranks of the observations in each sample and write them (R) in the next line under n.
4. Square the sum of ranks and write these (R^2) in a line under R.
5. Divide each value of R^2 by its respective value of n. Write this (R^2/n) in the bottom line under R^2. Add up the separate values of R^2/n to obtain $\Sigma(R^2/n)$.

The results of Steps 1 to 5 are shown in Table 16.1.

Table 16.1 Number of orchids (rank score in brackets)

	A	B	C	D	
	27 (12)	48 (16)	11 (6)	44 (15)	
	14 (7)	18 (9½)	0 (1)	72 (19)	
	8 (4½)	32 (13)	3 (2)	81 (20)	
	18 (9½)	51 (17)	15 (8)	55 (18)	
	7 (3)	22 (11)	8 (4½)	39 (14)	
n	5	5	5	5	$N = 20$
R	36	66.5	21.5	86	
R^2	1296	4422.25	462.25	7396	
R^2/n	259.2	884.45	92.45	1479.2	$\Sigma(R^2/n) = 2715.3$

6. The test statistic, K, is obtained by multiplying $\Sigma(R^2/n)$ by a factor $\dfrac{12}{N(N+1)}$ and then subtracting $3(N+1)$ where the numbers 12 and 3 are constants peculiar to this formula:

$$K = \left[\Sigma(R^2/n) \times \frac{12}{N(N+1)} \right] - 3(N+1)$$

$$K = \left[2715.3 \times \frac{12}{20(21)} \right] - 3(21)$$

$$K = 14.58$$

7. Compare K with the tabulated distribution of χ^2 (Appendix 3). The degrees of freedom is the number of samples less one ($4-1 = 3$ in this example). At 3 df our calculated value of 14.58 exceeds the tabulated value of 11.34 at $P = 0.01$. We reject the Null Hypothesis and conclude that there is a highly significant difference between the average number of orchids in the fields.

   It should be remembered that the test is applied to the samples *as a group* and that we are confident only that there are differences within the group as a whole. We should therefore be cautious in making inferences about differences between particular pairs of samples, or between one sample and the others. However, it is safe to assume that *at least* there is a significant difference between the two samples which have the highest and lowest sum of ranks. Inspecting the table of data we note that these are sample D and sample C, respectively. We infer that these two, at least, are significantly different from each other.

16.6 Advice on using the Kruskal–Wallis test

1. Apply the test to compare the locations (averages) of three or more samples. If there are only three samples there should be more than five observations in each sample.
2. The test statistic, K, is compared with the distribution of χ^2. This does not

mean however that observations have to be frequencies. Data may be on any scale of measurement that allows ranking and need not be normally distributed.

3. If the outcome of the test suggests rejection of H_0, be cautious about making *a posteriori*, that is, unplanned comparisons between samples, other than between the two with the highest and lowest sum of ranks. Having said that, 'common sense' should be used. It may be perfectly obvious from an inspection of the distribution of ranks that, for example, a particular sample stands out from the remainder.

16.7   The Wilcoxon's test for matched pairs

The Wilcoxon's test for matched pairs is a simple non-parametric test for comparing the medians of two matched samples. It calls for the calculation of a test statistic T whose probability distribution is known. In that test, one observation in a matched pair is subtracted from the other. Observations must therefore be measured on an *interval* scale. It is not possible to use this test for ordinal measurements such as abundance scales.

Example 16.3

An ornithologist working at a south coast reed swamp wishes to know if the habitat is used by migrating reed warblers for 'fattening up' before taking off on migration. Birds arrive in numbers during August and stay at least until the end of September. Birds weighed in September seem to be heavier than those in August. To test the significance of the relationship one could compare the median of a sample weighed in August with that of a sample in September using the Mann–Whitney U-test. If it were possible however to recapture some of the birds weighed in August during September and reweigh them, the two weights from each bird would constitute a *matched pair*. As we noted in Section 16.2, it is then possible to conduct a more sensitive analysis with matched data than with unmatched data. The Wilcoxon's test is appropriate for this problem.

The recorded masses are shown in Table 16.2. The Null Hypothesis H_0 is that there is no difference in median masses between the two sets of data. The alternative hypothesis, H_1, is that there is a difference, but with no prediction which way that difference will lie (i.e. a two-tailed test).

If H_0 is true, we have two expectations, namely:

(i) the number of mass *gains* will be matched by a similar number of mass *losses*. If H_1 is true, there will be more of one than the other.

(ii) the *sizes* of any mass changes will be balanced evenly between gains and losses; if H_1 is true, there will be a tendency for the larger changes to be in one direction.

The Wilcoxon's test for matched pairs quantifies both the direction and

Table 16.2   Matched masses in two bird samples

Mass of bird weighed in August, Sample A (g)	Mass of same bird weighed in September, Sample B (g)
10.3	12.2
11.4	12.1
10.9	13.1
12.0	11.9
10.0	12.0
11.9	12.9
12.2	11.4
12.3	12.1
11.7	13.5
12.0	12.3

magnitude of all the changes in a set of matched pairs. The steps in carrying out the test are as follows:

1. For each matched pair, subtract the value of observation A from the value of observation B. The answer is the difference, d. The value of d will have a negative sign if A is larger than B.
2. Rank the values of d according to their absolute values. That is to say, ignore the plus and minus signs for the moment. Ignore any instances in which $d=0$. If any ranks are tied, assign the average of the ranks exactly as described in Section 3.6.
3. Assign to each rank a '+' or a '−' sign corresponding to the sign of d.
4. Sum the ranks of the plus values and the minus values of d separately. The results of Steps 1–4 are given in Table 16.3.
   Sum of minus ranks: $1+5+2=8$
   Sum of plus ranks: $8+4+10+9+6+7+3=47$
   The smaller value of the two sums of ranks is the test statistic T.
   In this example, $T=8$.
5. Consult the table of the probability distribution of T (Appendix 7). When T is *equal to or less than* the critical value in the table, the Null Hypothesis is rejected at the particular level of significance. Enter the table at the appropriate value of N. N is not necessarily the total number of pairs of data, but the number of pairs less the number of pairs, for which $d=0$. Since there are none in the present example, $N=10$. Entering Appendix 7 at $N=10$, we find that our calculated value of T happens to be equal to the critical value of 8 at $P=0.05$ for a two-tailed test. The difference is therefore statistically significant.
6. Record the result of the test as 'there is a significant difference between the median masses of the two samples ($T=8$, $P<0.05$, Wilcoxon's test for matched pairs)'.

Table 16.3   Ranking of matched pairs

Sample A	Sample B	d	Rank of d
10.3	12.2	+1.9	+8
11.4	12.1	+0.7	+4
10.9	13.1	+2.2	+10
12.0	11.9	-0.1	-1
10.0	12.0	+2.0	+9
11.9	12.9	+1.0	+6
12.2	11.4	-0.8	-5
12.3	12.1	-0.2	-2
11.7	13.5	+1.8	+7
12.0	12.3	+0.3	+3

16.8   Advice on using the Wilcoxon's test for matched pairs

The Wilcoxon's test may only be applied when the value of one observation in a matched pair can be subtracted from the other. That is to say, they should be interval measurements, or counts of things. The number of matched pairs whose difference is not zero should be six or more. If the number of matched pairs exceeds about 40 the test is cumbersome and a parametric alternative (Section 16.13) is more appropriate.

1. Note that for H_0 to be rejected, T has to be *smaller* than or *equal to* the tabulated value at a given probability level.
2. The test is for a difference in *medians*. Do not be tempted to make statements about sample *means*.
3. The test assumes that samples have been drawn from parent populations which are symmetrically but not necessarily normally distributed. Because it may be impossible to discern the shape of the distribution in small samples, the point is largely academic. Nevertheless, if it is known from other studies that the two samples have been drawn from populations which have fundamentally asymmetrical distributions, do not use the test.

16.9   Comparing means – parametric tests

Parametric tests that compare means are more restrictive than their non-parametric counterparts. First, data should be recorded on interval or ratio scales of measurement. Second, data should be approximately normally distributed; that they are so can be checked as described in Section 9.6. It is possible to transform certain data, for example counts of things and proportions, to normal by means of an appropriate transformation (see Chapter 10). A third restriction is that the populations from which samples are drawn should have similar variances. The Null Hypothesis in a test for a difference between sample means is therefore:

H_0: two samples are drawn from populations with identical means and variances

It follows that if the outcome of a test suggests the rejection of H_0, we should eliminate the possibility that it is due to a difference between *variances* rather than between means. There is a simple test (the *F*-test) which decides if the difference between two sample variances is so small that it may be ignored. This check should be applied routinely before testing the difference between means, for if the outcome of an *F*-test suggests that variances are *not* similar, then a test for difference between means cannot be validly applied.

16.10 The *F*-test (two-tailed) = Variance ratio.

Two populations which have identical variances also have, by definition, a variance ratio of unity (that is, 1.00). Small samples drawn from such populations have a variance ratio which is distributed by sampling error around unity. At which point a departure from unity is so great that it cannot be accounted for by sampling error (and the populations from which the samples are drawn are presumed *not* to have equal variance) is decided by a **variance ratio**, or **F-test**. In that test the Null Hypothesis is:

$$H_0: \frac{\sigma_1^2}{\sigma_2^2} = F = 1.$$

A significant departure from unity is checked in tables of the distribution of F at the appropriate degrees of freedom. Because tabulated critical values of F are greater than 1, the greater variance is divided by the lesser variance:

$$F = \frac{\text{greater variance (Sample 1)}}{\text{lesser variance (Sample 2)}}$$

where the degrees of freedom (v) are ($n_1 - 1$) and ($n_2 - 1$) for Samples 1 and 2, respectively. v_1 v_2

Example 16.4

Two samples of limpets are collected randomly on either side of an island and the maximum shell length (mm) of each is measured. The sample statistics are:

Sample 1	*Sample 2*
West side	East side
$n_1 = 41$	$n_2 = 31$
$\bar{x}_1 = 24.71$	$\bar{x}_2 = 19.60$
$s_1 = 6.34$	$s_2 = 4.82$
$s_1^2 = 40.1956$	$s_2^2 = 23.2324$

Before commencing a test for a significant difference between *means* it is advisable first to check that there is no significant difference between *variances*. Substituting in the formula for F:

$$F = \frac{\text{greater variance}}{\text{lesser variance}} = \frac{40.1956}{23.2324} = 1.730$$

Consulting the distribution of F at $P = 0.05$ in Appendix 8 (two-tailed test because before the samples are drawn it is not known which has the larger variance), we find the critical value at $v_1 = 40$ and $v_2 = 30$ is 2.01. Our calculated value is below this. We therefore accept that the samples have been drawn from populations with equal or very similar variances. We may proceed to test for a difference between means with the assurance that a significant outcome will reflect a difference between means, not variances.

In our example we chose 41 and 31 observations so that the number of degrees of freedom are conveniently 40 and 30, respectively. The value of F is tabulated exactly at these numbers. Intermediate values of F are estimated by interpolation in the table.

16.11   The z-test for comparing the means of two large samples

When two populations have identical means then $\mu_1 = \mu_2$ and the population mean difference $(\mu_1 - \mu_2) = 0$. The value of the difference between the means of two samples drawn from these populations can be expressed as a deviation from the population mean difference, namely $(\bar{x}_1 - \bar{x}_2) - 0$. When the deviation is divided by the *standard error of the difference* it is transformed to a **z-score** in the same way that a single observation is transformed to a z-score when its deviation from the mean is divided by the standard deviation (Section 9.3). Inserting the expression for standard error of the difference (Section 11.6), and noting that subtracting zero is inconsequential and need not be written in, the full expression for z becomes:

$$z = \frac{(\bar{x}_1 - \bar{x}_2)}{\sqrt{\dfrac{s_1^2}{n_1} + \dfrac{s_2^2}{n_2}}}$$

Thus, we can transform the mean difference of any two samples to a z-score. As we know from Section 9.3 the critical values of z are 1.96 and 2.58. When a calculated value of z exceeds these, H_0 is rejected at $P = 0.05$ or $P = 0.01$, respectively, and we conclude that the samples are unlikely to have been drawn from populations with identical means.

From the Central Limit Theorem (Section 11.2) we know that the *means* of a series of samples drawn from a single population are normally distributed. It follows then that in applying the z-test, the populations from which samples are drawn do *not* have to be normally distributed provided that the samples are quite large (over 30 observations). In cases where the populations are suspected to be badly skewed, the sample sizes should exceed 50.

Example 16.5

Is it likely that the two limpet samples described in Example 16.4 are drawn from populations with identical means?

Substituting the data into the formula for z, we obtain:

$$z = \frac{(24.71 - 19.60)}{\sqrt{\left(\frac{40.1956}{41}\right) + \left(\frac{23.2324}{31}\right)}} = \frac{5.11}{\sqrt{(0.9804 + 0.7494)}} = \frac{5.11}{1.315}$$

$$z = 3.89$$

Our calculated value of z exceeds the value of 2.58 which corresponds to $P = 0.01$. The difference between sample means is highly significant. We therefore reject H_0 in favour of the alternative hypothesis that the samples are drawn from populations which have different means. It does not matter which sample is nominated Sample 1; in a two-tailed test a negative sign is ignored.

If there are independent grounds for predicting that one sample is drawn from a population that has a larger mean than the other, then a *one-tailed test* is called for. In a one-tailed test, the sample with the larger predicted mean is nominated Sample 1, in which case z will be positive only if H_1 is true. The critical value of z is lower in a one-tailed test, namely 1.65 at $P = 0.05$. We reiterate our earlier recommendation. Use the more stringent two-tailed test unless there is a clear-cut case for doing otherwise.

Readers should note that the value of 1.315 in the final step in the calculation of z above is the *standard error of the difference*. This is itself a useful number and its application was described in Section 11.6.

6.12 The *t*-test for comparing the means of two small samples

When samples are small (under about 30 observations in each) a different version of the z-test, called the **t-test**, is called for. The rationale is the same, namely that the mean difference between two samples is divided by the *standard error of the difference*. The answer is compared with the distribution of t at the appropriate degrees of freedom. As we explained in Section 11.7, the calculation of the standard error of the difference is a little more complicated in the case of small samples. Although the full formula for t appears rather cumbersome, all the terms within it are familiar, being the samples sizes, means and standard deviations:

$$t = \frac{(\bar{x}_1 - \bar{x}_2)}{\sqrt{\left[\frac{(n_1 - 1)s_1^2 + (n_2 - 1)s_2^2}{(n_1 + n_2 - 2)}\right]\left(\frac{n_1 + n_2}{n_1 n_2}\right)}}$$

where the degrees of freedom are $(n_1 + n_2) - 2$.

Unlike the z-test, the t-test *does* assume that samples have been drawn from populations which are normally distributed.

Example 16.6

A summary of sample data for measurements of males and females of a species is given below. Is it likely that they are drawn from populations with equal means; that is to say, are the sample means statistically significantly different?

Male (sample 1)   *Female (sample 2)*
$n_1 = 6$ $n_2 = 8$
$\bar{x}_1 = 74.8$ $\bar{x}_2 = 72.99$
$s_1 = 1.04$ $s_2 = 1.48$
$s_1^2 = 1.08$ $s_2^2 = 2.20$

First, a preliminary check to establish that variances are similar. Proceeding as in Example 16.4, we determine F to be $2.20/1.08 = 2.04$. Checking the distribution of F at $P = 0.05$ (two-tailed test) in Appendix 8, the tabulated F value at 7 and 5 df respectively is 6.85. The calculated value of F is well below this and we proceed with the t-test.

Substituting in the formula for t:

$$t = \frac{(74.8 - 72.99)}{\sqrt{\left[\frac{(6-1)1.08 + (8-1)2.20}{(6+8-2)}\right] \times \left(\frac{6+8}{6 \times 8}\right)}}$$

$$= \frac{1.81}{\sqrt{\left[\frac{5.4 + 15.4}{12}\right] \times 0.2917}}$$

$$= \frac{1.81}{\sqrt{1.733 \times 0.2917}} = \frac{1.81}{0.711}$$

$$= 2.55$$

(Note that the value of 0.711 in the final step is the *standard error of the difference*.)

Consulting the table of the distribution of t (Appendix 2) we find that the calculated value of 2.55 exceeds the tabulated value of 2.179 at $P = 0.05$ for $(14-2) = 12$ df. We reject the Null Hypothesis and conclude there is a statistically significant difference between the means.

The t-test may be applied as a *one-tailed test* exactly as described for the z-test. The distribution of t in a one-tailed test is given in Appendix 2.

16.13   The *t*-test for matched pairs

A specific form of the t-test is used when measurements constitute a matched pair. As noted in Section 16.2, it is possible to conduct a more sensitive test upon matched data. The rationale for the t-test with matched data is similar to that for the z- and t-tests with unmatched data. As we explained in Section 16.7 in connection with the Wilcoxon's test for matched pairs, it is possible to derive a difference, d, by subtracting the value of one observation in a matched pair from the other. If the Null Hypothesis that samples are drawn from populations with equal means is true, we would expect the mean value of d to be zero. We would not expect *all* values of d to be zero; rather, that all values would be normally distributed about the mean of zero. The standard deviation of the normal distribution is called the standard error of d. t is calculated by dividing the mean of the sample differences by the standard error of d. The overall formula for t is:

$$t = \frac{\Sigma d}{\sqrt{\dfrac{n\Sigma d^2 - (\Sigma d)^2}{(n-1)}}}$$

where n is the number of matched pairs and $(n-1)$ is the number of degrees of freedom.

Example 16.7

The reed warbler mass data employed to illustrate the Wilcoxon's test for matched pairs (Section 16.7) is used to demonstrate the procedure for computing the t-test for matched pairs. We assume that the Null Hypothesis is that there is no difference between the means of the samples and that H_1 is that there *is* a difference, and calls for a two-tailed test.

The table of data is reproduced in Table 16.4 and contains a column for the difference (d) between the values of each pair, and another for the squares of the

Table 16.4 t-test for matched pairs

Sample A mass in August (g)	Sample B mass in September (g)	d	d^2
10.3	12.2	+1.9	3.61
11.4	12.1	+0.7	0.49
10.9	13.1	+2.2	4.84
12.0	11.9	−0.1	0.01
10.0	12.0	+2.0	4.0
11.9	12.9	+1.0	1.0
12.2	11.4	−0.8	0.64
12.3	12.1	−0.2	0.04
11.7	13.5	+1.8	3.24
12.0	12.3	+0.3	0.09
		$\Sigma d = 8.8$	$\Sigma d^2 = 17.96$

difference (d^2). Values in these columns are then summed to give Σd and Σd^2 respectively. Notice that the sign of d is taken into account when summing the column and that the eventual value of t is not affected by which sample is nominated A or B.

From Σd we calculate $(\Sigma d)^2$ to be $8.8^2 = 77.44$. Substituting in the formula for t, with $n = 10$,

$$t = \frac{8.8}{\sqrt{\dfrac{(10 \times 17.96) - 77.44}{10 - 1}}} = \frac{8.8}{3.369} = 2.612$$

$t = 2.612$, at 9 degrees of freedom

Consulting a table of the distribution of t (Appendix 2) we find that for a two-tailed test our calculated value of t at 9 df exceeds the tabulated value of 2.262 at $P = 0.05$. We therefore reject the Null Hypothesis and conclude that the means of the two sets are significantly different in the two-tailed test.

16.14   Advice on comparing means

1. If there are more than 30 observations in each sample then use the z-test. If the distribution of data appears to be badly skewed, increase the number of observations to 50. If this is not possible, try transforming the data or use the Mann–Whitney U-test.
2. If there are fewer than 30 observations in each sample, use the t-test. Unlike the z-test, the t-test *does* assume that data are derived from normally distributed populations. Badly skewed data can sometimes be 'normalized' by transforming (see Chapter 10). Conduct the test on the transformed observations but do not back-transform the final value of t. If the data defy all attempts to normalize them, use the Mann–Whitney U-test.
3. Both z- and t-tests require that the variances of the two samples are similar. Check that they are so by means of a two-tailed F-test before conducting a z- or t-test.
4. z-tests or t-tests only answer the question: 'Is there a statistically significant difference between the means of two samples?' They do not address the more interesting question: '*To what extent* are the means different?'. The execution of both z- and t-tests requires the intermediate calculation of the *standard error of the difference*. This statistic may be employed to estimate the difference between the means of the populations from which the samples are drawn (Section 11.6).
5. A test for the difference between the means of two samples can also be carried out with a *one-way analysis of variance*. This is the subject of Chapter 17.

7 ANALYSIS OF VARIANCE – ANOVA

7.1 Why do we need ANOVA? *To make my life hell!*

Chapter 16 discusses ways of comparing the means of two samples. Sometimes however biologists wish to compare the means of more than two samples. Suppose, for example, we have length measurements from samples of three races of a species each of which lives on different islands A, B and C. It is possible to compare mean lengths by z-tests or, if the samples are small, by t-tests. We would need to perform the test three times to compare A–B, A–C and B–C. With the help of a calculator the task is not too daunting. Let us imagine instead that we wish to compare the means of seven samples. In this event no less than 21 z-tests are required to compare all possible pairs of means. Even if the analyst has sufficient patience to work through this cumbersome treatment, there is an underlying statistical objection to doing so.

We pointed out in Section 12.7 that if the $P = 0.05$ (5%) level of significance is consistently accepted, a wrong conclusion will be drawn on average once in every 20 tests performed. If the means of our hypothetical seven samples are compared in 21 z-tests, there is a good chance that at least one false conclusion will be drawn. Of course the risk of committing a Type 1 error, that is, rejecting H_0 when it should be accepted, is reduced by setting the acceptable significance level to the more stringent one of $P = 0.01$ (1%). But that increases the risk of making a Type 2 error, namely failing to reject H_0 when it should be rejected. **Analysis of variance (ANOVA)** overcomes these difficulties by allowing comparisons to be made between any number of sample means, all in a single test. When it is used in this way to compare the means of several samples statisticians speak of a one-way ANOVA.

ANOVA is such a flexible technique that it may also be used to compare more than one set of means. Referring to our island races again, it is possible to compare at the same time the mean lengths of samples of males and females of each race obtained from the islands. When the influence of two variables upon a sample mean is being analysed, such as island of origin and sex in our hypothetical example, the technique involved is described as a two-way ANOVA. Three-way, four-way (and so on) treatments are also possible but they get progressively more complicated. We restrict our examples in this text to one-way and two-way treatments.

17.2　How ANOVA works

How analysis of *variance* is used to investigate differences between *means* is illustrated in the following example.

Example 17.1

Compare the individual variances of the three samples below with the overall variance when all 15 observations $n=15$ are aggregated.

Sample 1	Sample 2	Sample 3	Overall
8	9	3	
10	11	5	
12	13	7	
14	15	9	
16	17	11	
$\Sigma x = 60$	$\Sigma x = 65$	$\Sigma x = 35$	$\Sigma x = 160$
$\bar{x} = 12.0$	$\bar{x} = 13.0$	$\bar{x} = 7.00$	$\bar{x} = 10.667$
$s^2 = 10.00$	$s^2 = 10.00$	$s^2 = 10.00$	$s^2 = 16.0$

The means of Samples 1 and 2 are similar; the mean of Sample 3 is much lower; the mean of the aggregated observations is intermediate in value. The variances of the three samples are identical (10.00) and therefore the 'average variance' is 10.00. The variance of the aggregated observations however is larger (16.0) than the average sample variance. The increase is due to the difference between the *means* of the samples, in particular, the difference between the mean of Sample 3 and the other two means. The samples thus give rise to two sources of variability:

(i) the variability around each mean *within* a sample (random scatter);
(ii) the variability *between* the samples due to differences between the means of the populations from which the samples are drawn.

In other words:

$$\text{Variability}_{total} = \text{variability}_{within} + \text{variability}_{between}$$

ANOVA involves the dividing up or **partitioning** the total variability of a number of samples into its components. If the samples are drawn from normally distributed populations with equal means and variances, the *within* variance is the same as the *between* variance. If a statistical test shows that this is not the case, then the samples have been drawn from populations with different means and/or variances. If it is assumed that the variances are equal (and this is an underlying assumption in ANOVA) then it is concluded that the discrepancy is due to differences between *means*. Thus:

$H_0 =$ samples are drawn from normally distributed populations with equal means and variances.

H_1 = population variances are assumed to be equal and therefore samples are drawn from populations with different means.

As we shall explain, the assumption that population variances are equal is not to be taken for granted; there is a simple check that should be made before ANOVA is applied. If it is suspected that observations are not approximately normally distributed (e.g. if they are counts) then ANOVA is performed upon *transformed* observations (see Chapter 10).

When partitioning the total variability of a number of samples, it is simpler to work with *sums of squares* because adding and subtracting variances is complicated by varying degrees of freedom. However, in the final stages of the analysis the sums of squares are converted to variances by dividing by the degrees of freedom in order to apply the F-test to compare them.

In Section 6.7 we gave the formula for estimating *sums of squares* (SS) as:

$$SS = \Sigma x^2 - \frac{(\Sigma x)^2}{n}$$

The quantity $\frac{(\Sigma x)^2}{n}$ is often referred to as the **correction term** (CT).

7.3   Procedure for computing one-way ANOVA

A biologist wishes to know if the mean masses of starlings sampled in four different roost situations are different. A sample of 10 units (starlings) is obtained from each situation. The procedure for conducting a one-way ANOVA is set out as a series of instructions.

1. Cast the data into a table, labelling each Sample 1–4, respectively. Use a scientific calculator to obtain, for each sample, the mean; the standard deviation (square this to obtain the variance); Σx (square this to obtain $(\Sigma x)^2$); and Σx^2. Record this information at the bottom of the column for each sample. At the right-hand side of the table, record the sums of the totals of n, Σx and Σx^2, using the subscript T to distinguish them from the sample data. These data are presented in Table 17.1.
2. Before proceeding with the main part of the analysis it is necessary to check that all four sample variances are similar to each other. This is called a test for the **homogeneity of variance**; it is undertaken by means of the F_{max} test. Only one test is required. If the largest and smallest variances of the samples are not significantly different from each other, then the others cannot be.

    Select the largest sample variance in the table and divide it by the smallest. Equate the results to F:

    $$F_{max} = \frac{17.14}{6.25} = 2.74, \text{ with 9 df in each sample}$$

Table 17.1 Masses of starlings from four roost situations (g)

Situation 1 Sample 1	Situation 2 Sample 2	Situation 3 Sample 3	Situation 4 Sample 4	Total
78	78	79	77	
88	78	73	69	
87	83	79	75	
88	81	75	70	
83	78	77	74	
82	81	78	83	
81	81	80	80	
80	82	78	75	
80	76	83	76	
89	76	84	75	
$n = 10$	$n = 10$	$n = 10$	$n = 10$	$n_T = 40$
$\bar{x} = 83.6$	$\bar{x} = 79.4$	$\bar{x} = 78.6$	$\bar{x} = 75.4$	
$s = 4.03$	$s = 2.50$	$s = 3.31$	$s = 4.14$	
$s^2 = 16.27$	$s^2 = 6.25$	$s^2 = 10.96$	$s^2 = 17.14$	
$\Sigma x = 836$	$\Sigma x = 794$	$\Sigma x = 786$	$\Sigma x = 754$	$\Sigma x_T = 3170$
$(\Sigma x)^2 = 698\,896$	$(\Sigma x)^2 = 630\,436$	$(\Sigma x)^2 = 617\,796$	$(\Sigma x)^2 = 568\,516$	
$\Sigma x^2 = 70\,036$	$\Sigma x^2 = 63\,100$	$\Sigma x^2 = 61\,878$	$\Sigma x^2 = 57\,006$	$\Sigma x_T^2 = 252\,020$

Consulting a table of the probability distribution of F_{max} (Appendix 9) we find that our calculated value of F is less than the critical value of 6.31 for number of samples $a = 4$ and df $(n-1) = 9$.

We conclude that the variances are homogeneous and we proceed with ANOVA.

The next five steps involve the partitioning of the sums of squares into the categories which make up the total. In each case the subscripts 1–4 pertain to data from Samples 1–4, respectively.

3. Calculate a factor called the correction term, CT:

$$CT = \frac{(\Sigma x_T)^2}{n_T} = \frac{(3170)^2}{40} = 251\,222.5$$

4. Calculate the total sum of squares of the aggregated samples, SS_T:

$$SS_T = \Sigma x_T^2 - CT$$
$$SS_T = 252\,020 - 251\,222.5 = 797.5$$

5. Calculate the *between samples* sum of squares, $SS_{between}$

$$SS_{between} = \frac{(\Sigma x_1)^2}{n_1} + \frac{(\Sigma x_2)^2}{n_2} + \frac{(\Sigma x_3)^2}{n_3} + \frac{(\Sigma x_4)^2}{n_4} - CT$$

$$SS_{between} = \frac{698\,896}{10} + \frac{630\,436}{10} + \frac{617\,796}{10} + \frac{568\,516}{10} - CT$$

$$SS_{between} = 69\,889.6 + 63\,043.6 + 61\,779.6 + 56\,851.6 - 251\,222.5$$
$$SS_{between} = 341.9$$

6. We now need to know the *within samples* sum of squares. Since we already know (Section 17.2) that

$$SS_T = SS_{between} + SS_{within}$$

we can derive SS_{within} simply by subtracting $SS_{between}$ from SS_T:

$$SS_{within} = (SS_T - SS_{between}) = (797.5 - 341.9) = 455.6$$

Whilst this 'short cut' method gives the correct result we advise SS_{within} be calculated independently because, if the sum of $SS_{between}$ and SS_{within} then checks with SS_T, we can be assured that no mistake has been made in the calculation. SS_{within} is the sum of the individual SS_{within} for each sample. We designate the individual values as SS_W.

$$SS_{W_1} = \Sigma x_1^2 - \frac{(\Sigma x_1)^2}{n_1} = 70\,036 - \frac{698\,896}{10} = 146.4 \qquad 4121 \cdot 71$$

$$SS_{W_2} = \Sigma x_2^2 - \frac{(\Sigma x_2)^2}{n_2} = 63\,100 - \frac{630\,436}{10} = 56.4 \qquad 38\,16 \cdot 86$$

$$SS_{W_3} = \Sigma x_3^2 - \frac{(\Sigma x_3)^2}{n_3} = 61\,878 - \frac{617\,796}{10} = 98.4$$

$$SS_{W_4} = \Sigma x_4^2 - \frac{(\Sigma x_4)^2}{n_4} = 57\,006 - \frac{568\,516}{10} = 154.4$$

$$SS_{within} = 146.4 + 56.4 + 98.4 + 154.4 = 455.6$$

7. Check that the independently calculated values of SS_{within} and $SS_{between}$ add up to that of SS_T calculated in Step 4:

$$455.6 + 341.9 = 797.5$$

8. Determine the number of degrees of freedom (df) for each of the calculated SS values. The rules for determining these are:

df for $SS_T = n_T - 1 = 40 - 1 = 39$
df for $SS_{between} = a - 1$ (where a = number of samples) $= (4 - 1) = 3$.
df for $SS_{within} = n_T - a = 40 - 4 = 36$

9. Estimate the variances by dividing each sum of squares by its respective degrees of freedom.

$$s^2_{between} = \frac{SS_{between}}{df_{between}} = \frac{341.9}{3} = 113.97$$

$$s^2_{within} = \frac{SS_{within}}{df_{within}} = \frac{455.6}{36} = 12.66$$

10. Compute F.

$$F = \frac{\text{Between samples variance}}{\text{Within samples variance}} = \frac{113.97}{12.66} = 9.002$$

It should be noted that the denominator, the bottom line in the division, is always the *within* samples variance. If it should turn out that this is larger than the *between* samples variance, then F is less than 1.0. This cannot be significant because tabulated values of F are greater than 1.0. Thus there is no need to compute F. H_0 is automatically accepted. Because a *nominated* variance (the *within* variance) is the denominator, this F test is *one-tailed* and the table in Appendix 10 is used.

11. Enter the result in an ANOVA summary table.

Source of variation	SS	df	s^2	F
Between	341.9	3	113.97	9.002
Within	455.6	36	12.66	
Total	797.5	39		

Consulting a table of the one-tailed distribution of F (Appendix 10), we find that our calculated value of F at 3 and 36 df exceeds the critical value of 2.88 (interpolating between 30 and 40 df in v_2). We therefore reject the Null Hypothesis, and conclude that the variation in the mean mass of the four starling samples are significantly different. We record the result as:

'The difference in mean mass of the four samples, where $n = 10$ in each case, is statistically significant ($F_{3,36} = 9.002$, $P < 0.05$).'

This is not the end of the analysis, however. Taking the four means as a group, we know that there is statistically significant variation between them. This may mean that all possible combinations of pairs are different from each other, or that just one is different from the other three. A good indication of which sample means are different from the others is obtained by presenting the individual sample means in histogram form, displaying the 95% intervals, exactly as in Fig. 11.2. The samples whose intervals do not overlap are presumed to have been drawn from populations with different means. We suggest you always do this because it indicates whether the outcome of your ANOVA is 'reasonable'.

A more sensitive test for distinguishing the mean differences which are significantly different is the **Tukey Test**. This is simple to apply and is outlined in the next section.

17.4 Procedure for computing the Tukey test

The Tukey test only needs to be undertaken when the result of the final F-test in the ANOVA indicates that there is a significant difference between the means of the groups. This version can only be used when there is an equal number of observations in all samples. The procedure for the test is as follows.

1. Construct a trellis for the comparison of all sample means. This is done below for the starling data used in the previous section. Any negative signs are ignored.

Sample	2	3	4
Sample 1 $\bar{x}_1 = 83.6$	$(\bar{x}_1 - \bar{x}_2)$ 4.2	$(\bar{x}_1 - \bar{x}_3)$ 5.0	$(\bar{x}_1 - \bar{x}_4)$ 8.2
Sample 2 $\bar{x}_2 = 79.4$		$(\bar{x}_2 - \bar{x}_3)$ 0.8	$(\bar{x}_2 - \bar{x}_4)$ 4.0
Sample 3 $\bar{x}_3 = 78.6$			$(\bar{x}_3 - \bar{x}_4)$ 3.2
Sample 4 $\bar{x}_4 = 75.4$			

2. Compute a test statistic T to provide a standard against which the values in the trellis will be tested.

$$T = (q) \times \sqrt{\frac{\text{within variance}}{n}}$$

where n = the number of sampling units in each sample. The *within variance* is obtained from the ANOVA summary table of Step 11 in the previous section. The value of q is found by consulting a table of the distribution of q (the Tukey Table, Appendix 11) for varying numbers of samples, a, and degrees of freedom, v. The respective values in this example are 4 and 36 (36 being the degrees of freedom of the within samples variance). Interpolating Appendix 11 at $a=4$, $v=36$, the tabulated value of q is about 3.82 (interpolating between 30 and 40 df).
   The T statistic is calculated as

$$T = 3.82 \times \sqrt{12.66/10} = 4.30$$

There are only two mean differences in the table which exceed this; 1 and 3, and 1 and 4. These means are therefore significantly different at $P=0.05$.
   The values of q in the Tukey Table have been calculated to take into account that several comparisons may be made and that there is a cumulative risk of committing a type 1 error. There is, of course, an associated increase in the risk of committing a type 2 error. Of the two, the latter is the more acceptable since it errs on the conservative side.

7.5   Two-way ANOVA

Two-way ANOVA techniques allow us to estimate the effects of two independent variables on a dependent variable. For example, a dependent variable might be mass; one of the independent variables might be sex, and season the other. Such data can be displayed in a contingency table similar to those described in Section 13.7. Instead of the data consisting of single counts, or frequencies, they represent the means of a number of observations. Hypothetical measurements for mean masses of both sexes of a species weighed in autumn and spring are set out in the table below.

		Variable B, sex	
		Male	*Female*
Variable A, Season	*Autumn*	Sample 1 57 g	Sample 2 53 g
	Spring	Sample 3 55 g	Sample 4 51 g

There are two particular questions that arise from the data:

(i) Is there a significant difference between the mean masses of males and females?
(ii) Is there a significant difference between the mean masses of birds in each season?

There are six combinations of pairs of means we need to compare in order to answer these questions: Autumn males–Autumn females; Autumn males–Spring males; Autumn males–Spring females; Autumn females–Spring males; Autumn females–Spring females; Spring males–Spring females. Whilst it is certainly possible to compare these means by z or t-tests, the same objections that are raised in Section 17.1 apply also here. There is, however, an additional reason why these tests are inadequate: z and t-tests fail to reveal any effect due to *interaction* of the two variables.

The idea of interaction is explained graphically. In Fig. 17.1a four sample means are plotted with hypothetical confidence intervals. Sample means of each sex are joined by lines which are roughly parallel. These indicate that the effects of the two variables (sex and season) are *additive*. That is to say, transition from Autumn to Spring has added (or, in this example, subtracted – subtraction is just negative addition!) an equal amount from the mean mass of each sex. Alternatively, we may take the view that transition from female to male has added an equal amount to the mean mass of each season. There is therefore no interaction between sex and season in Fig. 17.1a.

In Figs 17.1b and 17.1c the situation is different. In Fig. 17.1b transition from Autumn to Spring results in a proportionately greater change to the mean mass of females, and the variables sex and season are interactive. In Fig. 17.1c there is also interaction, but of a different kind. Transition from Autumn to Spring results in a proportionately lesser change to females than males.

If interaction is observed then a biological explanation may be sought. Hypothetically, Fig. 17.1a might represent the mean mass of non-breeding

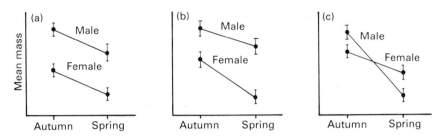

Fig. 17.1 Mean masses of samples of male and female animals weighed in Autumn and Spring: (a) no interaction; (b) positive interaction; (c) negative interaction. Confidence intervals are hypothetical.

males and females in each season. In Fig. 17.1b we might suppose that the Spring females were weighed just after egg-laying or birth of young when body reserves are depleted. Figure 17.1c might reflect Spring females which were weighed just prior to egg-laying or birth of young.

If interaction between variables is suspected, plot out the means as shown in Figs 17.1 and 17.2.

If there are more than two categories of each variable then the joined lines may form zig-zags. If the zig-zags are roughly parallel, there is no interaction; if the zig-zags are obviously not parallel then there may be interaction (Figs 17.2a, b).

Two-way ANOVA involves partitioning the total sums of squares of the samples into the various components that makes up the total variability. Once again there are two major components of the total sum of squares related in the equation:

$$SS_{total} = SS_{between} + SS_{within}$$

The SS_{within} quantity again represents the variability caused by random

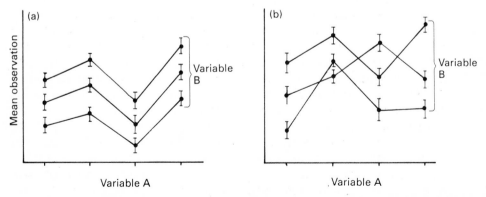

Fig. 17.2 Interactive effects between variables: (a) no interaction; (b) interaction. Confidence intervals are hypothetical.

effects within each sample. In the two-way ANOVA, the $SS_{between}$ item is subdivided into three components:

(i) SS representing variation between samples due to variable A (SS_A)
(ii) SS representing variation between samples due to variable B (SS_B)
(iii) SS representing variability between samples due to the interaction of variable A and variable B (SS_i).

The complete equation therefore may be written:

$$SS_T = (SS_A + SS_B + SS_i) + SS_{within}$$

In a two-way ANOVA we estimate all the components. The analysis concludes with separate comparisons of SS_A, SS_B and SS_i, with SS_{within}. As before, the individual values of SS are converted to variances by dividing by the appropriate degree of freedom so that F-tests may be used to check for significant differences between them.

17.6   Procedure for computing two-way ANOVA

Suppose that the starling data used to illustrate the one-way ANOVA procedure in Section 17.3 had been obtained in November. The biologist catches more birds in January in the same four roosting situations, and wishes to know not only if there are differences between the roosting situations, but also if there has been a change in mass over the winter. He also wishes to discover if there are any interactive effects between roosting situation and date of sampling. Two-way ANOVA is appropriate to the problem. The stepwise procedure is set out below.

1. Cast the measurements of each sample into a contingency table (in this case a 4×2 table), labelling each sample 1–8. Use a scientific calculator to obtain, for each sample, the mean; the standard deviation (square this to obtain s^2); Σx (square this to obtain $(\Sigma x)^2$); and Σx^2. Record this information with the sampling data in each cell. At the side of each row, and at the foot of each column, record the sums of the total of n, Σx and Σx^2 for each row and column, respectively, using the suffix t to distinguish them from the sample data. In the bottom right-hand cell of the table there is space to record the totals of these items for all rows (which is equal to the total for all columns) and represents the grand total. Use the suffix T to identify these. These data are presented in Table 17.2.
2. Homogeneity of variance check. Select the largest sample variance from the table, divide it by the smallest, and equate the result to F_{max} (as shown in Section 17.3, Step 2). In this example F_{max} is computed as $29.92/10.24 = 2.92$, for $a = 8$ and df $= 9$. This is less than the tabulated critical value of 8.95 (Appendix 9) and so we proceed with ANOVA.
3. Calculate the correction term, CT:

$$CT = \frac{(\Sigma x_T)^2}{n_T} = \frac{(6704)^2}{80} = 561\,795.2$$

Table 17.2 Two-way ANOVA of starling masses

Variable B – Roosting situation

Variable A – Month	Situation 1 / Column 1	Situation 2 / Column 2	Situation 3 / Column 3	Situation 4 / Column 4	Total for row
November Row 1	Sample 1 78 82 88 81 87 80 88 80 83 89 $n=10$; $\bar{x}=83.6$; $s^2=16.24$ $s=4.03$; $\Sigma x=836$ $(\Sigma x)^2=698\,896$ $\Sigma x^2=70\,036$	Sample 2 78 81 78 81 85 82 81 76 78 74 $n=10$; $\bar{x}=79.4$; $s^2=10.24$ $s=3.20$; $\Sigma x=794$ $(\Sigma x)^2=630\,436$ $\Sigma x^2=63\,136$	Sample 3 79 78 73 80 79 78 75 83 77 84 $n=10$; $\bar{x}=78.6$; $s^2=10.96$ $s=3.31$; $\Sigma x=786$ $(\Sigma x)^2=617\,796$ $\Sigma x^2=61\,878$	Sample 4 77 84 68 80 75 75 70 76 74 75 $n=10$; $\bar{x}=75.4$; $s^2=20.52$ $s=4.53$; $\Sigma x=754$ $(\Sigma x)^2=568\,516$ $\Sigma x^2=57\,036$	$n_t=40$ $\Sigma x_t=3170$ $\Sigma x_t^2=252\,086$
January Row 2	Sample 5 85 87 88 98 86 86 95 89 100 94 $n=10$; $\bar{x}=90.8$; $s^2=29.92$ $s=5.47$; $\Sigma x=908$ $(\Sigma x)^2=824\,464$ $\Sigma x^2=82\,716$	Sample 6 84 87 88 93 91 87 96 94 86 96 $n=10$; $\bar{x}=90.2$; $s^2=19.10$ $s=4.37$; $\Sigma x=902$ $(\Sigma x)^2=813\,604$ $\Sigma x^2=81\,532$	Sample 7 91 88 90 92 87 96 84 83 86 85 $n=10$; $\bar{x}=88.2$; $s^2=16.40$ $s=4.05$; $\Sigma x=882$ $(\Sigma x)^2=777\,924$ $\Sigma x^2=77\,940$	Sample 8 90 86 87 82 85 80 81 90 84 77 $n=10$; $\bar{x}=84.2$; $s^2=18.15$ $s=4.26$; $\Sigma x=842$ $(\Sigma x^2)=708\,964$ $\Sigma x^2=71\,060$	$n_t=40$ $\Sigma x_t=3534$ $\Sigma x_t^2=313\,248$
Total for column	$n_t=20$ $\Sigma x_t=1744$ $\Sigma x_t^2=152\,752$	$n_t=20$ $\Sigma x_t=1696$ $\Sigma x_t^2=144\,668$	$n_t=20$ $\Sigma x_t=1668$ $\Sigma x_t^2=139\,818$	$n_t=20$ $\Sigma x_t=1596$ $\Sigma x_t^2=128\,096$	$n_T=80$ $\Sigma x_T=6704$ $\Sigma x_T^2=565\,334$

4. Calculate the total sum of squares of the aggregated samples, SS_T

$$SS_T = \Sigma x_T^2 - CT$$
$$SS_T = 565\,334 - 561\,795.2 = 3538.8$$

5. Calculate the *between samples* sum of squares, $SS_{between}$

$$SS_{between} = \frac{(\Sigma x_1)^2}{n_1} + \frac{(\Sigma x_2)^2}{n_2} + \ldots \frac{(\Sigma x_8)^2}{n_8} - CT$$

where the suffixes 1–8 pertain to data from samples 1–8 respectively.

$$SS_{between} = \frac{(836)^2}{10} + \frac{(794)^2}{10} + \frac{(786)^2}{10} + \frac{(754)^2}{10} + \frac{(908)^2}{10} +$$

$$= \frac{(902)^2}{10} + \frac{(882)^2}{10} + \frac{(842)^2}{10} - CT$$

$$SS_{between} = 564\,060 - 561\,795.2$$
$$SS_{between} = 2264.8$$

6. Calculate the sum of squares for variable A.

$$SS_A = \frac{(\Sigma x_t \text{ row } 1)^2}{n_t \text{ row } 1} + \frac{(\Sigma x_t \text{ row } 2)^2}{n_t \text{ row } 2} - CT$$

$$SS_A = \frac{(3170)^2}{40} + \frac{(3534)^2}{40} - 561\,795.2$$

$$SS_A = 563\,451.4 - 561\,795.2$$
$$SS_A = 1656.2$$

7. Calculate the sum of squares for variable B, SS_B.

$$SS_B = \frac{(\Sigma x_t \text{ column } 1)^2}{n_t \text{ column } 1} + \frac{(\Sigma x_t \text{ column } 2)^2}{n_t \text{ column } 2} + \frac{(\Sigma x_t \text{ column } 3)^2}{n_t \text{ column } 3} +$$

$$\frac{(\Sigma x_t \text{ column } 4)^2}{n_t \text{ column } 4} - CT$$

$$SS_B = \frac{(1744)^2}{20} + \frac{(1696)^2}{20} + \frac{(1668)^2}{20} + \frac{(1596)^2}{20} - 561\,795.2$$

$$SS_B = 574.4$$

8. Calculate the sum of squares for the interaction, SS_i.

$$SS_i = SS_{between} - (SS_A + SS_B)$$
$$SS_i = 2264.8 - (1656.2 + 574.4)$$
$$SS_i = 34.2$$

9. Calculate the within sum of squares, SS_{within}.

$$SS_{within} = SS_T - SS_{between}$$
$$SS_{within} = 3538.8 - 2264.8$$
$$SS_{within} = 1274$$

Note that, as in Section 17.3, Step 6, it is possible to derive SS_{within} independently from SS_T and $SS_{between}$. This serves to confirm the accuracy of the calculations.

$$SS_{within} = \Sigma x_1^2 - \frac{(\Sigma x_1)^2}{n_1} + \ldots \Sigma x_8^2 - \frac{(\Sigma x_8)^2}{n_8}$$

10. Determine the degrees of freedom for each sum of squares. The rules are:

df for $SS_T = (n_T - 1) = 80 - 1 = 79$
df for $SS_{between} = (a - 1) = 8 - 1 = 7$ (where a = number of samples)
df for $SS_A = (r - 1) = 2 - 1 = 1$ (where r = number of rows)
df for $SS_B = (c - 1) = 4 - 1 = 3$ (where c = number of columns)
df for $SS_i = (r - 1)(c - 1) = 1 \times 3 = 3$
df for $SS_{within} = (n_T - rc) = 80 - (2 \times 4) = 72$

11. Estimate the variances by dividing all the sums of squares by their respective degrees of freedom.

$$Variance_T = \frac{SS_T}{df_T} = \frac{3538.8}{79} = 44.79$$

$$Variance_A = \frac{SS_A}{df_A} = \frac{1656.2}{1} = 1656.2$$

$$Variance_B = \frac{SS_B}{df_B} = \frac{574.4}{3} = 191.5$$

$$Variance_i = \frac{SS_i}{df_i} = \frac{34.2}{3} = 11.4$$

$$Variance_{within} = \frac{SS_{within}}{df_{within}} = \frac{1274}{72} = 17.69$$

12. Calculate the F values for the main effects.

$$F \text{ (variable A)} = \frac{Variance_A}{Variance_{within}} = \frac{1656.2}{17.69} = 93.62$$

$$F \text{ (variable B)} = \frac{Variance_B}{Variance_{within}} = \frac{191.5}{17.69} = 10.83$$

$$F \text{ (interaction)} = \frac{Variance_i}{Variance_{within}} = \frac{11.4}{17.69} = 0.64$$

13. Summarize the data in an ANOVA table:

Source of variation	Sum of squares	df	Variance	F
(Between samples)	(2264.8)	(7)		
Variable A	1656.2	1	1656.2	93.62**
Variable B	574.4	3	191.5	10.83**
Interaction	34.2	3	11.4	0.64
Within samples	1274	72	17.69	

** Significant at $P < 0.01$.

14. Refer to a table of the distribution of F (Appendix 10). Two of our three F values exceed the critical value at $P = 0.01$ for the appropriate numbers of degrees of freedom. Note that because the within-samples variance is larger than the interaction variance there is no real need to compute F for interaction.

   Our first Null Hypothesis is that there are no significant differences between the masses of starlings in different roosting situations (variable B). Our value of $F = 10.83$ exceeds that tabulated at $P = 0.01$, namely about 4.0 at df 3,72. We therefore reject H_0 and conclude that roosting situation *does* affect the mean weight.

   Our second Null Hypothesis is that there are no significant differences between the masses of starlings captured on the two sampling dates. Our value of $F = 93.62$ greatly exceeds the tabulated value at $P = 0.01$ of approximately 6.9 for df 1,72. We therefore reject H_0 and conclude that sampling date *does* affect the weight of starlings.

   Our third Null Hypothesis is that there is no interaction between roosting situation and sampling date which influences the mean masses. Our calculated value of $F = 0.64$ does not allow us to reject the Null Hypothesis. We therefore conclude that there is no interactive effect. As is the case in our example of one-way ANOVA, this is not necessarily the end of the analysis. Although we know that there are statistically significant differences between the means of the samples, we do not know *which particular means* are significantly different. As we explained at the end of Section 17.3, plotting the means of each sample with their 95% confidence intervals will reveal obvious differences between samples. However, the **Tukey Test** is more sensitive for pin-pointing differences between means.

17.7 Procedure for computing the Tukey Test in two-way ANOVA

The procedure is exactly the same as described in Section 17.4 for one-way ANOVA.

Construct a trellis for the comparison of all sample means. Table 17.3 refers to the two-way ANOVA data from Table 17.2. Any negative signs are ignored.

Table 17.3   Tukey trellis for two-way ANOVA

Sample	2	3	4	5	6	7	8
Sample 1 $\bar{x}=83.6$	$\bar{x}_1-\bar{x}_2$ 4.2	$\bar{x}_1-\bar{x}_3$ 5.0	$\bar{x}_1-\bar{x}_4$ 8.2	$\bar{x}_1-\bar{x}_5$ 7.2	$\bar{x}_1-\bar{x}_6$ 6.6	$\bar{x}_1-\bar{x}_7$ 4.6	$\bar{x}_1-\bar{x}_8$ 0.6
Sample 2 $\bar{x}=79.4$		$\bar{x}_2-\bar{x}_3$ 0.8	$\bar{x}_2-\bar{x}_4$ 4.0	$\bar{x}_2-\bar{x}_5$ 11.4	$\bar{x}_2-\bar{x}_6$ 10.8	$\bar{x}_2-\bar{x}_7$ 8.8	$\bar{x}_2-\bar{x}_8$ 4.8
Sample 3 $\bar{x}=78.6$			$\bar{x}_3-\bar{x}_4$ 3.2	$\bar{x}_3-\bar{x}_5$ 12.2	$\bar{x}_3-\bar{x}_6$ 11.6	$\bar{x}_3-\bar{x}_7$ 9.6	$\bar{x}_3-\bar{x}_8$ 5.6
Sample 4 $\bar{x}=75.4$				$\bar{x}_4-\bar{x}_5$ 15.4	$\bar{x}_4-\bar{x}_6$ 14.8	$\bar{x}_4-\bar{x}_7$ 12.8	$\bar{x}_4-\bar{x}_8$ 8.8
Sample 5 $\bar{x}=90.8$					$\bar{x}_5-\bar{x}_6$ 0.6	$\bar{x}_5-\bar{x}_7$ 2.6	$\bar{x}_5-\bar{x}_8$ 6.6
Sample 6 $\bar{x}=90.2$						$\bar{x}_6-\bar{x}_7$ 2.0	$\bar{x}_6-\bar{x}_8$ 6.0
Sample 7 $\bar{x}=88.2$							$\bar{x}_7-\bar{x}_8$ 4.0
Sample 8 $\bar{x}=84.2$							

Compute the test statistic from

$$T=(q)\sqrt{\frac{\text{within variance}}{n}}$$

where q is obtained from the Tukey Table (Appendix 11) as described in Section 17.4, and n is the number of observations in each sample. The values of a and v for these data are 8 and 72 respectively, giving us a tabulated value of q of approximately 4.4 (interpolating between 60 and 120 for v).

$$T=4.4\sqrt{\frac{17.69}{10}}$$

$$T=5.85$$

There are 15 out of a possible 28 pairs of means whose differences exceed this value and whose differences are, therefore, statistically significant.

17.8   Two-way ANOVA with single observations

In our example of two-way ANOVA described in Section 17.6 the data in each cell of the table consist of several (10) observations. It is not necessary to have several observations in each cell, but if there is only one observation we are unable to compute a *between samples* sum of squares (Step 5). Without this quantity it is impossible to derive a sum of squares for *interaction* and this source of variability cannot be investigated. In order to apply two-way ANOVA to single observations we are obliged to assume there is *no* interaction.

Example 17.2

A large nature reserve has a mosaic of habitats including willow scrub, birch scrub, reed swamp and grassland. Each year the owner recruits conservation volunteers to create new ponds throughout the reserve. A biologist investigates productivity in the ponds by collecting copepods and other planktonic invertebrates by means of standardized open-water sweeps with a plankton net. One-, two- and three-year-old ponds are selected at random in each of four habitats, thus generating 12 sampling units. Net contents are dried and weighed to produce 12 observations of biomass. They are displayed in Table 17.4, together with the mean, Σx and Σx^2 of each row and column. The bottom right cell gives the totals, identified by subscript T. The objective is to discover if there are significant differences between row means, due to pond age, or between column means, due to habitat. Our Null Hypothesis supposes that there are *no* differences due to these factors.

Table 17.4   Biomass of pond invertebrates (g)

		Variable B – Habitat				
		Column 1 Willow	Column 2 Birch	Column 3 Reed	Column 4 Grass	
Variable A – Age	Row 1 Year 1	3.0	2.7	4.5	1.5	$\bar{x}=2.925$ $\Sigma x=11.7$ $\Sigma x^2=38.79$
	Row 2 Year 2	3.3	4.2	6.3	3.7	$\bar{x}=4.375$ $\Sigma x=17.50$ $\Sigma x^2=81.91$
	Row 3 Year 3	5.2	6.8	9.7	4.7	$\bar{x}=6.60$ $\Sigma x=26.4$ $\Sigma x^2=189.46$
		$\bar{x}=3.833$ $\Sigma x=11.50$ $\Sigma x^2=46.93$	$\bar{x}=4.567$ $\Sigma x=13.7$ $\Sigma x^2=71.17$	$\bar{x}=6.833$ $\Sigma x=20.5$ $\Sigma x^2=154.03$	$\bar{x}=3.3$ $\Sigma x=9.9$ $\Sigma x^2=38.03$	$n_T=12$ $\Sigma x_T=55.6$ $\Sigma x^2_T=310.16$

In this example the total sum of squares SS_T is partitioned into three components:

$$SS_{Total} = SS_{Variable\ A} + SS_{Variable\ B} + SS_{Within}$$

The procedure is similar to that described in our previous examples.

1. Calculate CT

$$CT = \frac{(\Sigma x_T)^2}{n_T} = \frac{(55.6)^2}{12} = 257.613$$

2. Calculate the total sum of squares, SS_T

$$SS_T = \Sigma x^2{}_T - CT = (310.16 - 257.613) = 52.547$$

3. Calculate the sum of squares for variable A, SS_A

$$SS_A = \frac{(\Sigma x_1)^2}{n_c} + \frac{(\Sigma x_2)^2}{n_c} + \frac{(\Sigma x_3)^2}{n_c} - CT$$

where n_c is the number of columns (i.e. the number of observations in a row) and subscripts 1–3 pertain to rows 1–3.

$$SS_A = \frac{(11.7)^2}{4} + \frac{(17.5)^2}{4} + \frac{(26.4)^2}{4} - 257.613$$

$$SS_A = (285.025 - 257.613) = 27.412$$

4. Calculate the sum of squares for variable B, SS_B

$$SS_B = \frac{(\Sigma x_1)^2}{n_r} + \frac{(\Sigma x_2)^2}{n_r} + \frac{(\Sigma x_3)^2}{n_r} + \frac{(\Sigma x_4)^2}{n_r} - CT$$

where n_r is the number of rows (i.e. the number of observations in a column) and subscripts 1–4 pertain to columns 1–4.

$$SS_B = \frac{(11.5)^2}{3} + \frac{(13.7)^2}{3} + \frac{(20.5)^2}{3} + \frac{(9.9)^2}{3} - 257.613$$

$$SS_B = (279.4 - 257.613) = 21.787$$

5. Calculate the *within* sum of squares, SS_{within}

$$\begin{aligned}SS_{within} &= SS_T - (SS_A + SS_B) \\ &= 52.547 - (27.412 + 21.787) \\ &= 3.348\end{aligned}$$

6. Determine the degrees of freedom for each sum of squares. The rules are:

df for $SS_T = n_T - 1 = 11$
df for $SS_A = r - 1 = 2$
df for $SS_B = c - 1 = 3$
df for $SS_{within} = (r - 1)(c - 1) = 6$

7. Estimate the variances by dividing each sum of squares by its respective degrees of freedom:

$$s_T^2 = \frac{52.547}{11} = 4.777$$

$$s_A^2 = \frac{27.412}{2} = 13.706$$

$$s_B^2 = \frac{21.787}{3} = 7.262$$

$$s_{within}^2 = \frac{3.348}{6} = 0.558$$

8. Calculate F for each factor (variable):

$$F_{2,6(\text{Variable A})} = \frac{\text{Variance A}}{\text{Variance}_{within}} = \frac{13.706}{0.558} = 24.56$$

$$F_{3,6(\text{Variable B})} = \frac{\text{Variance B}}{\text{Variance}_{within}} = \frac{7.262}{0.558} = 13.01$$

9. Summarize the results in an ANOVA table:

Source of variation	Sum of squares	df	s^2	F
Variable A (age)	27.412	2	13.706	24.56**
Variable B (habitat)	21.787	3	7.262	13.01**
Within	3.348	6	0.558	
Total	52.547	11	4.777	

10. Refer to the distribution of F (Appendix 10). Both calculated values of F of 24.56 at 2,6 df for Variable A and 13.01 at 3,6 df exceed the respective tabulated critical values of 10.925 and 9.7795 at $P = 0.01$. We are obliged to reject the Null Hypotheses and conclude that the biomass of planktonic invertebrates is affected by the age of the pools and the habitats in which they exist. With only a single observation in each cell we are unable to apply the Tukey Test. However, inspection of the table of raw data suggests that biomass increases with age, is highest in the reed habitat and lowest in grassland.

In this example, our observations are on a continuous scale, namely *mass*. If observations are *counts* of organisms we can still proceed with ANOVA but, bearing in mind the conditions outlined in Section 10.1, data are usually first *transformed*. An example of ANOVA in which counts are transformed is described in Section 17.9.

17.9   The randomized block design

Sometimes biologists collect samples which appear suitable for analysis by one-way classification. If however the area over which the sampling units are obtained is rather large there may be an underlying, or *systematic*, source of variability due to a gradient in the environment. Such a gradient could be due to slope, drainage, exposure to a prevailing trend such as wind, illumination, pollution source, and so on. The variability contributes to the overall variability within the samples and reduces the sensitivity of an analysis. A version of two-way ANOVA may then be employed to measure the variability attributable to the environmental gradient. Its sum of squares is subtracted from the total sum of squares allowing a more sensitive analysis of the **main effect** – that is, the variability that the biologist is attempting to identify and measure.

Example 17.3

A biologist wishes to assess the effectiveness of different coloured water traps for catching hoverflies (Diptera: Syrphidae). Water traps are small plastic buckets whose interiors are painted to choice which contain a centimetre or two depth of water into which hoverflies are attracted. Four colours, brown, yellow, white and green are investigated – we call them *treatments*: A, B, C and D, respectively. Five traps of each colour are set out in a grid in a woodland in which there is a perceptible slope. The grid is arranged so that each row, or *block*, in the grid is at right angles to the slope. Each block contains one trap of each colour; the position of a particular colour trap is decided by reference to random numbers. The layout is shown in Table 17.5 and the numbers in **bold**

Table 17.5 Numbers of hoverflies captured

		\multicolumn{4}{c}{Treatment (main effect)}	Block totals (transformed observations)
	1	**4** A 0.602	**23** C 1.362	**9** B 0.954	**11** D 1.041	$\Sigma x = 3.959$ $\Sigma x^2 = 4.211$
	2	**11** B 1.041	**7** D 0.845	**9** A 0.954	**16** C 1.204	$\Sigma x = 4.044$ $\Sigma x^2 = 4.157$
Block	3	**14** B 1.146	**20** C 1.301	**5** D 0.699	**7** A 0.845	$\Sigma x = 3.991$ $\Sigma x^2 = 4.209$
	4	**15** A 1.176	**4** D 0.602	**19** C 1.279	**8** B 0.903	$\Sigma x = 3.960$ $\Sigma x^2 = 4.197$
	5	**8** D 0.903	**6** B 0.778	**21** C 1.322	**3** A 0.477	$\Sigma x = 3.480$ $\Sigma x^2 = 3.396$

Gradient

type are the numbers of hoverflies captured in each trap. Is there a variation in capture effectiveness due to the colour of trap? Is there a significant effect due to the slope in the environment?

Before commencing the analysis, we note two points. First, the five observations of each of the four treatments (colours) could be assembled into four samples. We could test for a difference between the mean number of hoverflies caught in each sample by a Kruskal–Wallis test (Section 16.5) or a one-way ANOVA. But then possible effects due to the slope are ignored. Second, the data consist of *counts*. Bearing Section 10.1 in mind, we should transform them before proceeding with a parametric test. This has the effect of normalizing the data, stabilizing the variances and, in some circumstances, helping to reduce error due to interaction. In this example it is easy to show that the variance of each treatment sample is greater than the mean. Thus a logarithmic transformation is suitable. Moreover, because there are no zero counts, the simple logarithm of each count is sufficient. The logarithm of each count is shown in *italic* type in the table. The actual counts are now ignored and the analysis is performed upon the transformed data. The procedure is similar to the ANOVA described in Section 17.8.

Sum of squares are partitioned as follows:

$$SS_{total} = SS_{main\ effect} + SS_{blocks} + SS_{within}$$

Block totals of Σx and Σx^2 of the logarithm transformed counts are shown in cells at the right of each block (row) of the table. Unlike Example 17.2, we do not sum the columns; instead, the five logarithm values for each treatment are summed separately:

Treatment A (Brown): $\Sigma x_1 = 4.054$ $\Sigma x_1^2 = 3.597$

Treatment B (Yellow): $\Sigma x_2 = 4.822$ $\Sigma x_2^2 = 4.728$

Treatment C (White): $\Sigma x_3 = 6.468$ $\Sigma x_3^2 = 8.381$

Treatment D (Green): $\Sigma x_4 = 4.09$ $\Sigma x_4^2 = 3.464$

The sums of n, x and x^2 for the whole grid are:

$n_T = 20$; $\Sigma x_T = 19.434$; $\Sigma x_T^2 = 20.17$

1. Calculate CT

$$CT = \frac{(\Sigma x_T)^2}{n_T} = \frac{(19.434)^2}{20} = 18.884$$

2. Calculate the *total* sum of squares, SS_T

$$SS_T = \Sigma x_T^2 - CT = (20.17 - 18.884) = 1.286$$

3. Calculate the sum of squares due to the main effect, that is, treatments (colours A–D), SS_M:

$$SS_M = \frac{(\Sigma x_1)^2}{n_B} + \frac{(\Sigma x_2)^2}{n_B} + \frac{(\Sigma x_3)^2}{n_B} + \frac{(\Sigma x_4)^2}{n_B} - CT$$

where n_B is the number of blocks (i.e. the number of observations for each treatment) and subscripts 1–4 pertain to treatments A–D.

$$SS_M = \frac{(4.054)^2}{5} + \frac{(4.822)^2}{5} + \frac{(6.468)^2}{5} + \frac{(4.09)^2}{5} - 18.884$$

$$SS_M = (19.650 - 18.884) = 0.766$$

4. Calculate the sum of squares due to the gradient, that is, blocks 1–5, SS_B:

$$SS_B = \frac{(\Sigma x_1)^2}{n_M} + \frac{(\Sigma x_2)^2}{n_M} + \frac{(\Sigma x_3)^2}{n_M} + \frac{(\Sigma x_4)^2}{n_M} + \frac{(\Sigma x_5)^2}{n_M} - CT$$

where n_M is the number of treatments (i.e. the number of observations in a block) and subscripts 1–5 pertain to blocks 1–5.

$$SS_B = \frac{(3.959)^2}{4} + \frac{(4.044)^2}{4} + \frac{(3.991)^2}{4} + \frac{(3.960)^2}{4} + \frac{(3.480)^2}{4} - 18.884$$

$$SS_B = (18.937 - 18.884) = 0.053$$

5. Calculate the *within* sum of squares, SS_{within}

$$\begin{aligned}
SS_{within} &= SS_T - (SS_M + SS_B) \\
&= 1.286 - (0.766 + 0.053) \\
&= 0.467
\end{aligned}$$

6. Determine the degrees of freedom for each sum of squares. The rules are:

df for $SS_T = n_T - 1 = 19$
df for $SS_M = M - 1 = 3$ (M = number of treatments)
df for $SS_B = B - 1 = 4$ (B = number of blocks)
df for $SS_{within} = (M-1)(B-1) = 12$

7. Estimate the variances by dividing each sum of squares by its respective degrees of freedom:

$$s_T^2 = \frac{1.286}{19} = 0.0677$$

$$s_M^2 = \frac{0.766}{3} = 0.2553$$

$$s_B^2 = \frac{0.053}{4} = 0.01325$$

$$s_{within}^2 = \frac{0.467}{12} = 0.0389$$

8. Calculate F for the main effect and the block effect:

$$F_{3,12(main\ effect)} = \frac{0.2553}{0.0389} = 6.563$$

$$F_{4,12(blocks)} = \frac{0.01325}{0.0389} = 0.3406$$

9. Summarize the results in an ANOVA table:

Source of variation	Sum of squares	df	s^2	F
Main effect (treatments)	0.766	3	0.2553	6.563**
Blocks (environmental gradient)	0.053	4	0.01325	0.3406
Within	0.467	12	0.0389	
Total	1.286	19	0.0677	

10. Refer to the distribution of *F* (Appendix 10). Our calculated value of *F* for the main effect (treatment) of 6.563 at 3,12 df exceeds the tabulated critical value of 5.9525 at $P = 0.01$. We therefore reject H_0 and conclude that there is a highly significant difference in the effectiveness of hoverfly traps of different colours. Inspection of the data in the table shows that Σx is the highest for white traps and lowest for brown traps. On the basis of this analysis the biologist may decide to use white water traps in studies of hoverflies. On the other hand, *F* for the blocks (gradient effect) is so small that it does not allow us to reject H_0. Variation due to the gradient is negligible.

11. In a situation like this where the *block effect* is negligible, a useful facility is permissible. We may choose to ignore its effect, recast the data into four samples corresponding to each treatment and apply a one-way ANOVA. Then, the *blocks* sum of squares are added to the *within* sum of squares $(0.053 + 0.467 = 0.52)$ and, likewise, the degrees of freedom are summed $(4 + 12 = 16)$. The *within* variance becomes $0.52/16 = 0.0325$ and the *F* value for the treatment is accordingly $0.2553/0.0325 = 7.855$. Notice that the value of *F* has increased. Clearly, in a marginal case, the increase may be sufficient to cross the critical threshold of statistical significance.

The randomized block design has many applications. In our example, observations are obtained from sampling units (traps) laid out in a grid on the ground where blocks comprise rows of traps. It is easy to imagine how the basic design can be adapted for observations obtained from a grid of quadrats, for example. Observations do not have to be obtained from a grid on the ground, however. A table of data like those in Example 17.3 could be obtained from five observers sampling four different sites; each observer generates a *block* of four observations, and each site is a *treatment*. The order in which sites are sampled by an observer is randomized. Similarly, a single observer obtaining observations of four treatments on five successive days could generate such blocks providing that the sequence of observation each day is randomized and the observations of each day comprise a block.

.10   The Latin square

In Section 17.9 we considered a situation in which an investigation of a main effect (treatments) is compounded by the possible systematic influence of an environmental gradient in one direction across the sampling area. By extension, we could encounter situations where there are systematic effects of *two* environmental gradients running approximately at right angles. For example, a meadow bounded on one side by a river might sustain a slope, drainage or water availability gradient towards the river; a motorway bounding a second side at right angles to the river might introduce a gradient of lead or other vehicle exhaust pollutants in a second direction, towards the motorway.

A version of ANOVA known as the **Latin square** design enables the partitioning of sums of squares due to observations of *treatments*, that is, the main effect the observer is interested in, and the sums of squares attributable to each of the gradients. The design is especially suitable for investigating the effects of different treatments such as fertilizers, management techniques, etc., on some variable which can be measured in quadrats placed on the ground. The quadrat grid has an equal number of rows and columns representing the number of different treatments and constructed so that a particular treatment occurs once only in each row and column.

A   B   C
B   C   A   **A Latin square**
C   A   B

Example 17.4

A conservationist wishes to assess the effectiveness of each of four possible management treatments for promoting representative species such as meadow brome *Bromus commutatus* in a hay meadow nature reserve. The treatments are:

A: Hay cut and harvested
B: Hay flail cut and left on ground in windrows
C: Hay scythed and left *in situ*
D: Hay not cut

The meadow slopes in one direction towards a river; on another side, at right angles, it is bounded by a motorway. Any investigation into the main effect or treatment must take into account possible systematic variability due to gradients introduced by the river and motorway.

A 4 × 4 Latin square of 16 quadrats is marked out, allowing each treatment to be replicated four times. The position of each treatment in the grid is decided randomly by drawing tokens out of a bag with the restriction that a particular treatment occurs once only in each row and column. The layout is shown in Table 17.6. In the year following the treatments, the proportional frequency of

Table 17.6 Proportional frequency of meadow brome

⟵─────────── Gradient B due to river ───────────⟶

		Column 1	Column 2	Column 3	Column 4	Row totals (transformed observations)
River / Row	1	**0.32** D *34.45*	**0.81** A *64.16*	**0.64** B *53.13*	**0.57** C *49.02*	$\Sigma x = 200.76$ $\Sigma x^2 = 10\,529.07$
	2	**0.84** A *66.42*	**0.27** D *31.31*	**0.58** C *49.60*	**0.62** B *51.94*	$\Sigma x = 199.27$ $\Sigma x^2 = 10\,549.86$
	3	**0.63** C *52.54*	**0.67** B *54.94*	**0.79** A *62.73*	**0.19** D *25.84*	$\Sigma x = 196.05$ $\Sigma x^2 = 10\,381.61$
	4	**0.72** B *58.05*	**0.65** C *53.73*	**0.24** D *29.33*	**0.70** A *56.79*	$\Sigma x = 197.90$ $\Sigma x^2 = 10\,342.07$
		$\Sigma x = 211.46$ $\Sigma x^2 =$ $11\,728.67$	$\Sigma x = 204.14$ $\Sigma x^2 =$ $11\,002.14$	$\Sigma x = 194.79$ $\Sigma x^2 =$ $10\,078.26$	$\Sigma x = 183.59$ $\Sigma x^2 =$ $8\,993.53$	$\Sigma n_T = 16$ $\Sigma x_T = 793.98$ $\Sigma x_T^2 = 41\,802.61$

(Right margin: Gradient A due to motorway)

Motorway

meadow brome is estimated in each quadrat of the Latin square and the values obtained are shown in **bold**. Is there a variation in the proportional frequency of meadow brome between the treatments? Is there a systematic variation due to either or both of the gradients?

Before commencing the analysis, we note that observations are *proportions*. Bearing Section 10.1 in mind, we should transform the observations before proceeding with a parametric test. The *arcsine transformation* is suitable for proportions. The arcsine of each observation, obtained as described in Section 10.5, is shown in *italics* in Table 17.6. The actual observations are now ignored and the analysis performed upon the transformed data. Notice also that due to the symmetry of the Latin square, the number of treatments n is equal to be number of rows and the number of columns (four in this example). Sums of squares are partitioned as follows, where the *main effect* refers to the management treatments:

$$SS_{Total} = SS_{main\ effect} + SS_{gradient\ A} + SS_{gradient\ B} + SS_{within}$$

Row totals and column totals of Σx and Σx^2 of the arcsine-transformed data are shown in cells at the right of each row and the foot of each column. The sums for the whole grid, distinguished by the subscript T, are shown in the bottom right-hand cell. Sums of the arcsine-transformed treatment observations are listed separately:

Treatment A (hay cut and harvested): $\Sigma x = 250.1$ $\Sigma x^2 = 15\,688.28$
Treatment B (hay flail cut): $\Sigma x = 218.06$ $\Sigma x^2 = 11\,908.77$
Treatment C (hay scythed): $\Sigma x = 204.89$ $\Sigma x^2 = 10\,510.48$
Treatment D (hay not cut): $\Sigma x = 120.93$ $\Sigma x^2 = 3695.073$

The procedure is as follows.

1. Calculate CT

$$CT = \frac{(\Sigma x_T)^2}{n_T} = \frac{(793.98)^2}{16} = 39\,400.27$$

2. Calculate the total sum of squares, SS_T

$$SS_T = \Sigma x_T^2 - CT = (41\,802.61 - 39\,400.27) = 2402.34$$

3. Calculate the sum of squares due to the *main effect*, that is, treatments A–D, SS_M:

$$SS_M = \frac{(\Sigma x_1)^2}{n} + \frac{(\Sigma x_2)^2}{n} + \frac{(\Sigma x_3)^2}{n} + \frac{(\Sigma x_4)^2}{n} - CT$$

where n is the number of treatments and subscripts 1–4 pertain to treatments A–D.

$$SS_M = \frac{(250.1)^2}{4} + \frac{(218.06)^2}{4} + \frac{(204.89)^2}{4} + \frac{(120.93)^2}{4} - 39\,400.27$$

$$= (41\,676.04 - 39\,400.27) = 2275.77$$

4. Calculate the sum of squares due to gradient A, distance from the motorway, SS_A

$$SS_A = \frac{(\Sigma x_1)^2}{n} + \frac{(\Sigma x_2)^2}{n} + \frac{(\Sigma x_3)^2}{n} + \frac{(\Sigma x_4)^2}{n} - CT$$

where $n =$ number of columns (i.e. the number of observations in a row) and subscripts 1–4 pertain to rows 1–4.

$$SS_A = \frac{(200.76)^2}{4} + \frac{(199.27)^2}{4} + \frac{(196.05)^2}{4} + \frac{(197.90)^2}{4} - 39\,400.27$$

$$= (39\,403.28 - 39\,400.27) = 3.01$$

5. Calculate the sum of squares due to gradient B, distance from the river, SS_B

$$SS_B = \frac{(\Sigma x_1)^2}{n} + \frac{(\Sigma x_2)^2}{n} + \frac{(\Sigma x_3)^2}{n} + \frac{(\Sigma x_4)^2}{n} - CT$$

where $n =$ the number of rows (i.e. the number of observations in a column) and subscripts 1–4 pertain to columns 1–4.

$$SS_B = \frac{(211.46)^2}{4} + \frac{(204.14)^2}{4} + \frac{(194.79)^2}{4} + \frac{(183.59)^2}{4} - CT$$

$$= (39\,509.23 - 39\,400.27) = 108.96$$

6. Calculate the *within* sum of squares, SS_{within}

$$SS_{within} = SS_T - (SS_M + SS_A + SS_B)$$
$$= 2402.34 - (2275.77 + 3.01 + 108.96) = 14.6$$

7. Determine the degrees of freedom for each sum of squares. The rules are:

df for $SS_T = n_T - 1 = 15$
df for $SS_M = n - 1 = 3$
df for $SS_A = n - 1 = 3$
df for $SS_B = n - 1 = 3$
df for $SS_{within} = (n-1)(n-2) = 6$

where n_T is the total number of observations and n is the number of treatments (= number of rows = number of columns).

8. Estimate the variances by dividing each sum of squares by its respective degrees of freedom.

$$s_T^2 = \frac{2402.34}{15} = 160.156$$

$$s_M^2 = \frac{2275.77}{3} = 758.59$$

$$s_A^2 = \frac{3.01}{3} = 1.003$$

$$s_B^2 = \frac{108.96}{3} = 36.32$$

$$s_{within}^2 = \frac{14.6}{6} = 2.433$$

9. Calculate F for the main (treatment) effect and the two gradient effects.

$$F_{3,6(treatments)} = \frac{758.59}{2.433} = 311.79$$

$$F_{3,6(gradient\ A)} = \frac{1.003}{2.433} = 0.412$$

$$F_{3,6(gradient\ B)} = \frac{36.32}{2.433} = 14.928$$

10. Summarize the results in an ANOVA table:

Source of variation	Sum of squares	df	s^2	F
Main effect (treatments)	2275.77	3	758.59	311.79**
Gradient A (distance from motorway)	3.01	3	1.003	0.412
Gradient B (distance from river)	108.96	3	36.32	14.928**
Within	14.6	6	2.433	
Total	2402.34	15	160.156	

11. Refer to the distribution of F (Appendix 10). Our calculated value of F for the main effect (treatment) of 311.79 at 3,6 df greatly exceeds the tabulated critical value of 9.7795 at $P=0.01$. We reject H_0 and conclude there is a statistically very highly significant difference between the effects of the treatments on the proportion of meadow brome. Inspecting the values of Σx for the treatments, we see that treatment A promotes the species whilst treatment D suppresses it.

   We note too that the calculated value of $F=14.928$ for gradient B, distance from the river, also exceeds the critical value. Inspecting the values of Σx for the columns we note that the proportion of the species declines with increasing distance from the river. Perhaps it likes damp conditions. Our calculated value of F for gradient A, distance from the motorway, is very small (<1) and does not allow us to reject H_0 for this effect. We infer that there is no pollution gradient from the motorway which affects the species.

12. As explained in Example 17.3, Step 11, if one (or both) of the gradient effects is negligible, its (their) sums of squares and degrees of freedom may be combined with the respective *within* values. This reduces the *within variance* and increases the value(s) of F for the remaining effect(s) which, in marginal cases, may make the difference between accepting or rejecting H_0.

11 Analysis of variance in regression

We said in Section 15.10 that analysis of variance may be used as an alternative to the *t*-test for testing the significance of a *least squares* regression line. In Section 15.15 we noted that ANOVA is the only suitable method for testing the significance of a regression line obtained by the *reduced major axis* method. The principle of ANOVA in regression is to determine the sum of squares due to the regression (*main effect*) and the residual sum of squares (*within*). Each is converted to a variance by dividing by the respective degrees of freedom, 1 and

$(n-2)$, respectively. To complete the test the regression variance is divided by the residual variance to obtain an F value which is compared with tabulated critical values of F at 1 $(n-1)$ df.

Example 17.5

Refer to the regression problem of Example 15.2. Establish the significance of the regression line by means of the F test in ANOVA. The quantities needed are derived in Section 15.8, page 151.

1. Calculate the regression sum of squares, $SS_{regression}$

$$SS_{regression} = \frac{(sum\ of\ products)^2}{sum\ of\ squares\ of\ x}$$

These quantities are worked out in Steps (a) and (c) in Section 15.8.

$$SS_{regression} = \frac{(41\,837.5)^2}{51\,562.5} = 33\,946.694$$

The sum of squares for regression always has 1 df. Therefore the regression variance is equal to the regression sum of squares.

2. Calculate the residual variance.
   This quantity is obtained in Step (d) of Section 15.8 and is 360.98 with $(n-2)$ df where n is the number of *pairs* of observations. In this example there are $(10-2) = 8$ df.

3. Calculate F by dividing the regression variance by the residual variance

$$F_{1,8} = \frac{33\,946.694}{360.98} = 94.040$$

4. Our calculated F value of 94.04 at 1,8 df greatly exceeds the tabulated critical value of 11.26 at these df at $P = 0.01$. We conclude that the regression is highly significant. The outcome is, of course, consistent with the result of the t-test performed on the same data in Section 15.10.

17.12   Advice on using ANOVA

ANOVA is an efficient and powerful technique for investigating relationships between samples. There are, however, a number of restrictions to the use of the technique.

1. ANOVA assumes that all observations are obtained randomly and are normally distributed. Certain data may be *transformed* to normal as described in Chapter 10.
2. ANOVA assumes that variances of the samples are similar. This can be checked with the F_{max} test. If the variances between samples are greatly different, transformation of data will usually stabilize the variances.

3. In performing the F test in ANOVA the *within* variance always appears in the denominator of the test. Because it is nominated as the smaller variance, this F test is *one-tailed*. Tables of F distribution (Appendix 10) are published for a one-tailed test. Do *not* confuse them with the two-tailed tables (Appendix 8) used for testing the equality of variance prior to a t-test.

4. In one-way ANOVA, sample sizes do not have to be equal. They do have to be equal, however, if the Tukey test is to be applied. In the example of two-way or interactive ANOVA we describe in Section 17.6, sample sizes must be equal. Field biologists are often in a position to ensure that this is the case by deciding how many quadrats, traps, and so on, are to be laid out. In Section 17.6, this is not the case however. It is rather unlikely that equal numbers of starlings will be captured in all situations. A practical expedient is to set the sample size to the smallest sample and select the same number of units to be measured randomly from the larger samples. This is admittedly wasteful, but the alternative is a complex analysis suitable for computer treatment and beyond the scope of this text.

5. In two-way ANOVA, interaction cannot be measured where data in each cell of a contingency table consist of single observations. Variability due to interaction is combined with the *within* variability and it is assumed to be negligible.

APPENDICES

Appendix 1: Table of random numbers

23157	54859	01837	25993	76249	70886	95230	36744
05545	55043	10537	43508	90611	83744	10962	21343
14871	60350	32404	36223	50051	00322	11543	80834
38976	74951	94051	75853	78805	90194	32428	71695
97312	61718	99755	30870	94251	25841	54882	10513
11742	69381	44339	30872	32797	33118	22647	06850
43361	28859	11016	45623	93009	00499	43640	74036
93806	20478	38268	04491	55751	18932	58475	52571
49540	13181	08429	84187	69538	29661	77738	09527
36768	72633	37948	21569	41959	68670	45274	83880
07092	52392	24627	12067	06558	45344	67338	45320
43310	01081	44863	80307	52555	16148	89742	94647
61570	06360	06173	63775	63148	95123	35017	46993
31352	83799	10779	18941	31579	76448	62584	86919
57048	86526	27795	93692	90529	56546	35065	32254
09243	44200	68721	07137	30729	75756	09298	27650
97957	35018	40894	88329	52230	82521	22532	61587
93732	59570	43781	98885	56671	66826	95996	44569
72621	11225	00922	68264	35666	59434	71687	58167
61020	74418	45371	20794	95917	37866	99536	19378
97839	85474	33055	91718	45473	54144	22034	23000
89160	97192	22232	90637	35055	45489	88438	16361
25966	88220	62871	79265	02823	52862	84919	54883
81443	31719	05049	54806	74690	07567	65017	16543
11322	54931	42362	34386	08624	97687	46245	23245

Appendix 2: *t*-distribution

	Level of significance for one-tailed test			
	0.05	0.025	0.01	0.005
	Level of significance for two-tailed test			
d.f.	0.10	0.05	0.02	0.01
1	6.314	12.706	31.821	63.657
2	2.920	4.303	6.965	9.925
3	2.353	3.182	4.541	5.841
4	2.132	2.776	3.747	4.604
5	2.015	2.571	3.365	4.032
6	1.943	2.447	3.143	3.707
7	1.895	2.365	2.998	3.499
8	1.860	2.306	2.896	3.355
9	1.833	2.262	2.821	3.250
10	1.812	2.228	2.764	3.169
11	1.796	2.201	2.718	3.106
12	1.782	2.179	2.681	3.055
13	1.771	2.160	2.650	3.012
14	1.761	2.145	2.624	2.977
15	1.753	2.131	2.602	2.947
16	1.746	2.120	2.583	2.921
17	1.740	2.110	2.567	2.898
18	1.734	2.101	2.552	2.878
19	1.729	2.093	2.539	2.861
20	1.725	2.086	2.528	2.845
21	1.721	2.080	2.518	2.831
22	1.717	2.074	2.508	2.819
23	1.714	2.069	2.500	2.807
24	1.711	2.064	2.492	2.797
25	1.708	2.060	2.485	2.787
26	1.706	2.056	2.479	2.779
27	1.703	2.052	2.473	2.771
28	1.701	2.048	2.467	2.763
29	1.699	2.045	2.462	2.756
30	1.697	2.042	2.457	2.750
40	1.684	2.021	2.423	2.704
60	1.671	2.000	2.390	2.660
120	1.658	1.980	2.358	2.617
∞	1.645	1.960	2.326	2.576

Appendix 3: χ^2 distribution

Degrees of freedom	Level of significance	
	0.05	0.01
1	3.84	6.63
2	5.99	9.21
3	7.81	11.34
4	9.49	13.28
5	11.07	15.09
6	12.59	16.81
7	14.07	18.48
8	15.51	20.09
9	16.92	21.67
10	18.31	23.21
11	19.68	24.72
12	21.03	26.22
13	22.36	27.69
14	23.68	29.14
15	25.00	30.58
16	26.30	32.00
17	27.59	33.41
18	28.87	34.81
19	30.14	36.19
20	31.41	37.57
21	32.67	38.93
22	33.92	40.29
23	35.17	41.64
24	36.42	42.98
25	37.65	44.31
26	38.89	45.64
27	40.11	46.96
28	41.34	48.28
29	42.56	49.59
30	43.77	50.89
40	55.76	63.69
50	67.50	76.15
60	79.08	88.38
70	90.53	100.43
80	101.88	112.33
90	113.15	124.12
100	124.34	135.81

Appendix 4: Critical values of Spearman's Rank Correlation Coefficient

	Level of significance for one-tailed test			
	0.05	0.025	0.01	0.005
	Level of significance for two-tailed test			
n	0.10	0.05	0.02	0.01
5	0.900	—	—	—
6	0.829	0.886	0.943	—
7	0.714	0.786	0.893	—
8	0.643	0.738	0.833	0.881
9	0.600	0.683	0.783	0.833
10	0.564	0.648	0.745	0.794
11	0.523	0.623	0.736	0.818
12	0.497	0.591	0.703	0.780
13	0.475	0.566	0.673	0.745
14	0.457	0.545	0.646	0.716
15	0.441	0.525	0.623	0.689
16	0.425	0.507	0.601	0.666
17	0.412	0.490	0.582	0.645
18	0.399	0.476	0.564	0.625
19	0.388	0.462	0.549	0.608
20	0.377	0.450	0.534	0.591
21	0.368	0.438	0.521	0.576
22	0.359	0.428	0.508	0.562
23	0.351	0.418	0.496	0.549
24	0.343	0.409	0.485	0.537
25	0.336	0.400	0.475	0.526
26	0.329	0.392	0.465	0.515
27	0.323	0.385	0.456	0.505
28	0.317	0.377	0.448	0.496
29	0.311	0.370	0.440	0.487
30	0.305	0.364	0.432	0.478

Appendix 5: Product moment correlation values at the 0.05 and 0.01 levels of significance

d.f.	0.05	0.01		d.f.	0.05	0.01
1	0.997	0.9999		35	0.325	0.418
2	0.950	0.990		36	0.320	0.413
3	0.878	0.959		38	0.312	0.403
4	0.811	0.917		40	0.304	0.393
5	0.754	0.874		42	0.297	0.384
6	0.707	0.834		44	0.291	0.376
7	0.666	0.798		45	0.288	0.372
8	0.632	0.765		46	0.284	0.368
9	0.602	0.735		48	0.279	0.361
10	0.576	0.708		50	0.273	0.354
11	0.553	0.684		55	0.261	0.338
12	0.532	0.661		60	0.250	0.325
13	0.514	0.641		65	0.241	0.313
14	0.497	0.623		70	0.232	0.302
15	0.482	0.606		75	0.224	0.292
16	0.468	0.590		80	0.217	0.283
17	0.456	0.575		85	0.211	0.275
18	0.444	0.561		90	0.205	0.267
19	0.433	0.549		95	0.200	0.260
20	0.423	0.537		100	0.195	0.254
21	0.413	0.526		125	0.174	0.228
22	0.404	0.515		150	0.159	0.208
23	0.396	0.505		175	0.148	0.193
24	0.388	0.496		200	0.138	0.181
25	0.381	0.487		300	0.113	0.148
26	0.374	0.479		400	0.098	0.128
27	0.367	0.471		500	0.088	0.115
28	0.361	0.463		1,000	0.062	0.081
29	0.355	0.456				
30	0.349	0.449				
32	0.339	0.436				
34	0.329	0.424				

Appendix 6: Mann–Whitney U-test Values (two-tailed test) $P = 0.05$

n_1 \ n_2	2	3	4	5	6	7	8	9	10	11	12	13	14	15	16	17	18	19	20
2							0	0	0	0	1	1	1	1	1	2	2	2	2
3				0	1	1	2	2	3	3	4	4	5	5	6	6	7	7	8
4			0	1	2	3	4	4	5	6	7	8	9	10	11	11	12	13	13
5		0	1	2	3	5	6	7	8	9	11	12	13	14	15	17	18	19	20
6		1	2	3	5	6	8	10	11	13	14	16	17	19	21	22	24	25	27
7		1	3	5	6	8	10	12	14	16	18	20	22	24	26	28	30	32	34
8	0	2	4	6	8	10	13	15	17	19	22	24	26	29	31	34	36	38	41
9	0	2	4	7	10	12	15	17	20	23	26	28	31	34	37	39	42	45	48
10	0	3	5	8	11	14	17	20	23	26	29	33	36	39	42	45	48	52	55
11	0	3	6	9	13	16	19	23	26	30	33	37	40	44	47	51	55	58	62
12	1	4	7	11	14	18	22	26	29	33	37	41	45	49	53	57	61	65	69
13	1	4	8	12	16	20	24	28	33	37	41	45	50	54	59	63	67	72	76
14	1	5	9	13	17	22	26	31	36	40	45	50	55	59	64	67	74	78	83
15	1	5	10	14	19	24	29	34	39	44	49	54	59	64	70	75	80	85	90
16	1	6	11	15	21	26	31	37	42	47	53	59	64	70	75	81	86	92	98
17	2	6	11	17	22	28	34	39	45	51	57	63	67	75	81	87	93	99	105
18	2	7	12	18	24	30	36	42	48	55	61	67	74	80	86	93	99	106	112
19	2	7	13	19	25	32	38	45	52	58	65	72	78	85	92	99	106	113	119
20	2	8	13	20	27	34	41	48	55	62	69	76	83	90	98	105	112	119	127

n_1 and n_2 are the number of observations in each sample

Appendix 7: Critical values of *T* in the Wilcoxon's test for two matched samples

Sample size	Levels of significance			
	One-tailed test			
	0.05	0.025	0.01	0.001
	Two-tailed test			
	0.1	0.05	0.02	0.002
N = 5	T ≤ 0			
6	2	0		
7	3	2	0	
8	5	3	1	
9	8	5	3	
10	10	8	5	0
11	13	10	7	1
12	17	13	9	2
13	21	17	12	4
14	25	21	15	6
15	30	25	19	8
16	35	29	23	11
17	41	34	27	14
18	47	40	32	18
19	53	46	37	21
20	60	52	43	26
21	67	58	49	30
22	75	65	55	35
23	83	73	62	40
24	91	81	69	45
25	100	89	76	51
26	110	98	84	58
27	119	107	92	64
28	130	116	101	71
30	151	137	120	86
31	163	147	130	94
32	175	159	140	103
33	187	170	151	112

Appendix 8: *F*-distribution, 0.05 level of significance, two-tailed test

v_2 \ v_1	1	2	3	4	5	6	7	8	9	10	12	15	20	24	30	40	60	120	∞
1	647.8	799.5	864.2	899.6	921.8	937.1	948.2	956.7	963.3	968.6	976.7	984.9	993.1	997.2	1001	1006	1010	1014	1018
2	38.51	39.00	39.17	39.25	39.30	39.33	39.36	39.37	39.39	39.40	39.41	39.43	39.45	39.46	39.46	39.47	39.48	39.49	39.50
3	17.44	16.04	15.44	15.10	14.88	14.73	14.62	14.54	14.47	14.42	14.34	14.25	14.17	14.12	14.08	14.04	13.99	13.95	13.90
4	12.22	10.65	9.98	9.60	9.36	9.20	9.07	8.98	8.90	8.84	8.75	8.66	8.56	8.51	8.46	8.41	8.36	8.31	8.26
5	10.01	8.43	7.76	7.39	7.15	6.98	6.85	6.76	6.68	6.62	6.52	6.43	6.33	6.28	6.23	6.18	6.12	6.07	6.02
6	8.81	7.26	6.60	6.23	5.99	5.82	5.70	5.60	5.52	5.46	5.37	5.27	5.17	5.12	5.07	5.01	4.96	4.90	4.85
7	8.07	6.54	5.89	5.52	5.29	5.12	4.99	4.90	4.82	4.76	4.67	4.57	4.47	4.42	4.36	4.31	4.25	4.20	4.14
8	7.57	6.06	5.42	5.05	4.82	4.65	4.53	4.43	4.36	4.30	4.20	4.10	4.00	3.95	3.89	3.84	3.78	3.73	3.67
9	7.21	5.71	5.08	4.72	4.48	4.32	4.20	4.10	4.03	3.96	3.87	3.77	3.67	3.61	3.56	3.51	3.45	3.39	3.33
10	6.94	5.46	4.83	4.47	4.24	4.07	3.95	3.85	3.78	3.72	3.62	3.52	3.42	3.37	3.31	3.26	3.20	3.14	3.08
11	6.72	5.26	4.63	4.28	4.04	3.88	3.76	3.66	3.59	3.53	3.43	3.33	3.23	3.17	3.12	3.06	3.00	2.94	2.88
12	6.55	5.10	4.47	4.12	3.89	3.73	3.61	3.51	3.44	3.37	3.28	3.18	3.07	3.02	2.96	2.91	2.85	2.79	2.72
13	6.41	4.97	4.35	4.00	3.77	3.60	3.48	3.39	3.31	3.25	3.15	3.05	2.95	2.89	2.84	2.78	2.72	2.66	2.60
14	6.30	4.86	4.24	3.89	3.66	3.50	3.38	3.29	3.21	3.15	3.05	2.95	2.84	2.79	2.73	2.67	2.61	2.55	2.49
15	6.20	4.77	4.15	3.80	3.58	3.41	3.29	3.20	3.12	3.06	2.96	2.86	2.76	2.70	2.64	2.59	2.52	2.46	2.40
16	6.12	4.69	4.08	3.73	3.50	3.34	3.22	3.12	3.05	2.99	2.89	2.79	2.68	2.63	2.57	2.51	2.45	2.38	2.32
17	6.04	4.62	4.01	3.66	3.44	3.28	3.16	3.06	2.98	2.92	2.82	2.72	2.62	2.56	2.50	2.44	2.38	2.32	2.25
18	5.98	4.56	3.95	3.61	3.38	3.22	3.10	3.01	2.93	2.87	2.77	2.67	2.56	2.50	2.44	2.38	2.32	2.26	2.19
19	5.92	4.51	3.90	3.56	3.33	3.17	3.05	2.96	2.88	2.82	2.72	2.62	2.51	2.45	2.39	2.33	2.27	2.20	2.13
20	5.87	4.46	3.86	3.51	3.29	3.13	3.01	2.91	2.84	2.77	2.68	2.57	2.46	2.41	2.35	2.29	2.22	2.16	2.09
21	5.83	4.42	3.82	3.48	3.25	3.09	2.97	2.87	2.80	2.73	2.64	2.53	2.42	2.37	2.31	2.25	2.18	2.11	2.04
22	5.79	4.38	3.78	3.44	3.22	3.05	2.93	2.84	2.76	2.70	2.60	2.50	2.39	2.33	2.27	2.21	2.14	2.08	2.00
23	5.75	4.35	3.75	3.41	3.18	3.02	2.90	2.81	2.73	2.67	2.57	2.47	2.36	2.30	2.24	2.18	2.11	2.04	1.97
24	5.72	4.32	3.72	3.38	3.15	2.99	2.87	2.78	2.70	2.64	2.54	2.44	2.33	2.27	2.21	2.15	2.08	2.01	1.94
25	5.69	4.29	3.69	3.35	3.13	2.97	2.85	2.75	2.68	2.61	2.51	2.41	2.30	2.24	2.18	2.12	2.05	1.98	1.91
26	5.66	4.27	3.67	3.33	3.10	2.94	2.82	2.73	2.65	2.59	2.49	2.39	2.28	2.22	2.16	2.09	2.03	1.95	1.88
27	5.63	4.24	3.65	3.31	3.08	2.92	2.80	2.71	2.63	2.57	2.47	2.36	2.25	2.19	2.13	2.07	2.00	1.93	1.85
28	5.61	4.22	3.63	3.29	3.06	2.90	2.78	2.69	2.61	2.55	2.45	2.34	2.23	2.17	2.11	2.05	1.98	1.91	1.83
29	5.59	4.20	3.61	3.27	3.04	2.88	2.76	2.67	2.59	2.53	2.43	2.32	2.21	2.15	2.09	2.03	1.96	1.89	1.81
30	5.57	4.18	3.59	3.25	3.03	2.87	2.75	2.65	2.57	2.51	2.41	2.31	2.20	2.14	2.07	2.01	1.94	1.87	1.79
40	5.42	4.05	3.46	3.13	2.90	2.74	2.62	2.53	2.45	2.39	2.29	2.18	2.07	2.01	1.94	1.88	1.80	1.72	1.64
60	5.29	3.93	3.34	3.01	2.79	2.63	2.51	2.41	2.33	2.27	2.17	2.06	1.94	1.88	1.82	1.74	1.67	1.58	1.48
120	5.15	3.80	3.23	2.89	2.67	2.52	2.39	2.30	2.22	2.16	2.05	1.94	1.82	1.76	1.69	1.61	1.53	1.43	1.31
∞	5.02	3.69	3.12	2.79	2.57	2.41	2.29	2.19	2.11	2.05	1.94	1.83	1.71	1.64	1.57	1.48	1.39	1.27	1.00

Use this table for checking equality of variance prior to a *z*-test or a *t*-test

v_1, v_2 are the degrees of freedom of the samples with larger and smaller variance, respectively.

Appendix 9: Critical values of F_{max} 0.05 level of significance

Use this table when checking homogeneity of variance preceding Analysis of Variance

v \ a	2	3	4·	5	6	7	8	9	10	11	12
2	39.0	87.5	142	202	266	333	403	475	550	626	704
3	15.4	27.8	39.2	50.7	62.0	72.9	83 5	93.9	104	114	124
4	9.60	15.5	20.6	25.2	29.5	33.6	37.5	41.1	44.6	48.0	51.4
5	7.15	10.8	13.7	16.3	18.7	20.8	22.9	24.7	26.5	28.2	29.9
6	5.82	8.38	10.4	12.1	13.7	15.0	16.3	17.5	18.6	19.7	20.7
7	4.99	6.94	8.44	9.70	10.8	11.8	12.7	13.5	14.3	15.1	15.8
8	4.43	6.00	7.18	8.12	9.03	9.78	10.5	11.1	11.7	12.2	12.7
9	4.03	5.34	6.31	7.11	7.80	8.41	8.95	9.45	9.91	10.3	10.7
10	3.72	4.85	5.67	6.34	6.92	7.42	7.87	8.28	8.66	9.01	9.34
12	3.28	4.16	4.79	5.30	5.72	6.09	6.42	6.72	7.00	7.25	7.48
15	2.86	3.54	4.01	4.37	4.68	4.95	5.19	5.40	5.59	5.77	5.93
20	2.46	2.95	3.29	3.54	3.76	3.94	4.10	4.24	4.37	4.49	4.59
30	2.07	2.40	2.61	2.78	2.91	3.02	3.12	3.21	3.29	3.36	3.39
60	1.67	1.85	1.96	2.04	2.11	2.17	2.22	2.26	2.30	2.33	2.36

a is the number of samples being compared. v is the degrees of freedom of each sample (if samples do not have equal numbers of observations then use the degrees of freedom of the sample with the smaller number of observations).

Appendix 10: *F*-distribution

Use these tables for testing significance in analysis of Variance.
(a) 0.05 level
(b) 0.01 level

$v_1 = $ df for the greater variance
$v_2 = $ df for the lesser variance

(a)

v_2 \ v_1	1	2	3	4	5	6	7	8	9
1	161.45	199.50	215.71	224.58	230.16	233.99	236.77	238.88	240.54
2	18.513	19.000	19.164	19.247	19.296	19.330	19.353	19.371	19.385
3	10.128	9.5521	9.2766	9.1172	9.0135	8.9406	8.8867	8.8452	8.8323
4	7.7086	6.9443	6.5914	6.3882	6.2561	6.1631	6.0942	6.0410	5.9938
5	6.6079	5.7861	5.4095	5.1922	5.0503	4.9503	4.8759	4.8183	4.7725
6	5.9874	5.1433	4.7571	4.5337	4.3874	4.2839	4.2067	4.1468	4.0990
7	5.5914	4.7374	4.3468	4.1203	3.9715	3.8660	3.7870	3.7257	3.6767
8	5.3177	4.4590	4.0662	3.8379	3.6875	3.5806	3.5005	3.4381	3.3881
9	5.1174	4.2565	3.8625	3.6331	3.4817	3.3738	3.2927	3.2296	3.1789
10	4.9646	4.1028	3.7083	3.4780	3.3258	3.2172	3.1355	3.0717	3.0204
11	4.8443	3.9823	3.5874	3.3567	3.2039	3.0946	3.0123	2.9480	2.8962
12	4.7472	3.8853	3.4903	3.2592	3.1059	2.9961	2.9134	2.8486	2.7964
13	4.6672	3.8056	3.4105	3.1791	3.0254	2.9153	2.8321	2.7669	2.7444
14	4.6001	3.7389	3.3439	3.1122	2.9582	2.8477	2.7642	2.6987	2.6458
15	4.5431	3.6823	3.2874	3.0556	2.9013	2.7905	2.7066	2.6408	2.5876
16	4.4940	3.6337	3.2389	3.0069	2.8524	2.7413	2.6572	2.5911	2.5377
17	4.4513	3.5915	3.1968	2.9647	2.8100	2.6987	2.6143	2.5480	2.4443
18	4.4139	3.5546	3.1599	2.9277	2.7729	2.6613	2.5767	2.5102	2.4563
19	4.3807	3.5219	3.1274	2.8951	2.7401	2.6283	2.5435	2.4768	2.4227
20	4.3512	3.4928	3.0984	2.8661	2.7109	2.5990	2.5140	2.4471	2.3928
21	4.3248	3.4668	3.0725	2.8401	2.6848	2.5727	2.4876	2.4205	2.3660
22	4.3009	3.4434	3.0491	2.8167	2.6613	2.5491	2.4638	2.3965	2.3219
23	4.2793	3.4221	3.0280	2.7955	2.6400	2.5277	2.4422	2.3748	2.3201
24	4.2597	3.4028	3.0088	2.7763	2.6207	2.5082	2.4226	2.3551	2.3002
25	4.2417	3.3852	2.9912	2.7587	2.6030	2.4904	2.4047	2.3371	2.2821
26	4.2252	3.3690	2.9752	2.7426	2.5868	2.4741	2.3883	2.3205	2.2655
27	4.2100	3.3541	2.9604	2.7278	2.5719	2.4591	2.3732	2.3053	2.2501
28	4.1960	3.3404	2.9467	2.7141	2.5581	2.4453	2.3593	2.2913	2.2360
29	4.1830	3.3277	2.9340	2.7014	2.5454	2.4324	2.3463	2.2783	2.2329
30	4.1709	3.3158	2.9223	2.6896	2.5336	2.4205	2.3343	2.2662	2.2507
40	4.0847	3.2317	2.8387	2.6060	2.4495	2.3359	2.2490	2.1802	2.1240
60	4.0012	3.1504	2.7581	2.5252	2.3683	2.2541	2.1665	2.0970	2.0401
120	3.9201	3.0718	2.6802	2.4472	2.2899	2.1750	2.0868	2.0164	1.9688
∞	3.8415	2.9957	2.6049	2.3719	2.2141	2.0986	2.0096	1.9384	1.8799

Appendix 10 (*cont.*)

(*a cont.*)

10	12	15	20	24	30	40	60	120	∞
241.88	243.91	245.95	248.01	249.05	250.10	251.14	252.20	253.25	254.31
19.396	19.413	19.429	19.446	19.454	19.462	19.471	19.479	19.487	19.496
8.7855	8.7446	8.7029	8.6602	8.6385	8.6166	8.5944	8.5720	8.5594	8.5264
5.9644	5.9117	5.8578	5.8025	5.7744	5.7459	5.7170	5.6877	5.6381	5.6281
4.7351	4.6777	4.6188	4.5581	4.5272	4.4957	4.4638	4.4314	4.3085	4.3650
4.0600	3.9999	3.9381	3.8742	3.8415	3.8082	3.7743	3.7398	3.7047	3.6689
3.6365	3.5747	3.5107	3.4445	3.4105	3.3758	3.3404	3.3043	3.2674	3.2298
3.3472	3.2839	3.2184	3.1503	3.1152	3.0794	3.0428	3.0053	2.9669	2.9276
3.1373	3.0729	3.0061	2.9365	2.9005	2.8637	2.8259	2.7872	2.7475	2.7067
2.9782	2.9130	2.8450	2.7740	2.7372	2.6996	2.6609	2.6211	2.5801	2.5379
2.8536	2.7876	2.7186	2.6464	2.6090	2.5705	2.5309	2.4901	2.4480	2.4045
2.7534	2.6866	2.6169	2.5436	2.5055	2.4663	2.4259	2.3842	2.3410	2.2962
2.6710	2.6037	2.5331	2.4589	2.4202	2.3803	2.3392	2.2966	2.2524	2.2064
2.6022	2.5342	2.4630	2.3879	2.3487	2.3082	2.2664	2.2229	2.1778	2.1307
2.5437	2.4753	2.4034	2.3275	2.2878	2.2468	2.2043	2.1601	2.1141	2.0658
2.4935	2.4247	2.3522	2.2756	2.2354	2.1938	2.1507	2.1058	2.0589	2.0096
2.4499	2.3807	2.3077	2.2304	2.1898	2.1477	2.1040	2.0584	2.0107	1.9604
2.4117	2.3421	2.2686	2.1906	2.1497	2.1071	2.0629	2.0166	1.9681	1.9168
2.3779	2.3080	2.2341	2.1555	2.1141	2.0712	2.0264	1.9795	1.9302	1.8780
2.3479	2.2776	2.2033	2.1242	2.0825	2.0391	1.9938	1.9464	1.8963	1.8432
2.3210	2.2504	2.1757	2.0960	2.0540	2.0102	1.9645	1.9165	1.8657	1.8117
2.2967	2.2258	2.1508	2.0707	2.0283	1.9842	1.9380	1.8894	1.8380	1.7831
2.2747	2.2036	2.1282	2.0476	2.0050	1.9605	1.9139	1.8648	1.8128	1.7570
2.2547	2.1834	2.1077	2.0267	1.9838	1.9390	1.8920	1.8424	1.7896	1.7330
2.2365	2.1649	2.0889	2.0075	1.9643	1.9192	1.8718	1.8217	1.7684	1.7110
2.2197	2.1479	2.0716	1.9898	1.9464	1.9010	1.8533	1.8027	1.7488	1.6906
2.2043	2.1323	2.0558	1.9736	1.9299	1.8842	1.8361	1.7851	1.7306	1.6717
2.1900	2.1179	2.0411	1.9586	1.9147	1.8687	1.8203	1.7689	1.7138	1.6541
2.1768	2.1045	2.0275	1.9446	1.9005	1.8543	1.8055	1.7537	1.6981	1.6376
2.1646	2.0921	2.0148	1.9317	1.8874	1.8409	1.7918	1.7396	1.6835	1.6223
2.0772	2.0035	1.9245	1.8389	1.7929	1.7444	1.6928	1.6373	1.5766	1.5089
1.9926	1.9174	1.8364	1.7480	1.7001	1.6491	1.5943	1.5343	1.4673	1.3893
1.9105	1.8337	1.7505	1.6587	1.6084	1.5543	1.4952	1.4290	1.3519	1.2539
1.8307	1.7522	1.6664	1.5705	1.5173	1.4591	1.3940	1.3180	1.0214	1.0000

Appendix 10 (*cont.*)

(*b*)

v_1 / v_2	1	2	3	4	5	6	7	8	9
1	4052.2	4999.5	5403.4	5624.6	5763.6	5859.0	5928.4	5981.1	6022.5
2	98.503	99.000	99.166	99.249	99.299	99.333	99.356	99.374	99.388
3	34.116	30.817	29.457	28.710	28.237	27.911	27.672	27.489	27.345
4	21.198	18.000	16.694	15.977	15.522	15.207	14.976	14.799	14.659
5	16.258	13.274	12.060	11.392	10.967	10.672	10.456	10.289	10.158
6	13.745	10.925	9.7795	9.1483	8.7459	8.4661	8.2600	8.1017	7.9761
7	12.246	9.5466	8.4513	7.8466	7.4604	7.1914	6.9928	6.8400	6.7188
8	11.259	8.6491	7.5910	7.0061	6.6318	6.3707	6.1776	6.0289	5.9106
9	10.561	8.0215	6.9919	6.4221	6.0569	5.8018	5.6129	5.4671	5.3511
10	10.044	7.5594	6.5523	5.9943	5.6363	5.3858	5.2001	5.0567	4.9424
11	9.6460	7.2057	6.2167	5.6683	5.3160	5.0692	4.8861	4.7445	4.6315
12	9.3302	6.9266	5.9525	5.4120	5.0643	4.8206	4.6395	4.4994	4.3875
13	9.0738	6.7010	5.7394	5.2053	4.8616	4.6204	4.4410	4.3021	4.1911
14	8.8616	6.5149	5.5639	5.0354	4.6950	4.4558	4.2779	4.1399	4.0297
15	8.6831	6.3589	5.4170	4.8932	4.5556	4.3183	4.1415	4.0045	3.8948
16	8.5310	6.2262	5.2922	4.7726	4.4374	4.2016	4.0259	3.8896	3.7804
17	8.3997	6.1121	5.1850	4.6690	4.3359	4.1015	3.9267	3.7910	3.6822
18	8.2854	6.0129	5.0919	4.5790	4.2479	4.0146	3.8406	3.7054	3.5971
19	8.1849	5.9259	5.0103	4.5003	4.1708	3.9386	3.7653	3.6305	3.5225
20	8.0960	5.8489	4.9382	4.4307	4.1027	3.8714	3.6987	3.5644	3.4567
21	8.0166	5.7804	4.8740	4.3688	4.0421	3.8117	3.6396	3.5056	3.3981
22	7.9454	5.7190	4.8166	4.3134	3.9880	3.7583	3.5867	3.4530	3.3458
23	7.8811	5.6637	4.7649	4.2636	3.9392	3.7102	3.5390	3.4057	3.2986
24	7.8229	5.6136	4.7181	4.2184	3.8951	3.6667	3.4959	3.3629	3.2560
25	7.7698	5.5680	4.6755	4.1774	3.8550	3.6272	3.4568	3.3239	3.2172
26	7.7213	5.5263	4.6366	4.1400	3.8183	3.5911	3.4210	3.2884	3.1818
27	7.6767	5.4881	4.6009	4.1056	3.7848	3.5580	3.3882	3.2558	3.1494
28	7.6356	5.4529	4.5681	4.0740	3.7539	3.5276	3.3581	3.2259	3.1195
29	7.5977	5.4204	4.5378	4.0449	3.7254	3.4995	3.3303	3.1982	3.0920
30	7.5625	5.3903	4.5097	4.0179	3.6990	3.4735	3.3045	3.1726	3.0665
40	7.3141	5.1785	4.3126	3.8283	3.5138	3.2910	3.1238	2.9930	2.8876
60	7.0771	4.9774	4.1259	3.6490	3.3389	3.1187	2.9530	2.8233	2.7185
120	6.8509	4.7865	3.9491	3.4795	3.1735	2.9559	2.7918	2.6629	2.5586
∞	6.6349	4.6052	3.7816	3.3192	3.0173	2.8020	2.6393	2.5113	2.4073

Appendix 10 (*cont.*)

(*b cont.*)

10	12	15	20	24	30	40	60	120	∞
6055.8	6106.3	6157.3	6208.7	6234.6	6260.6	6286.8	6313.0	6339.4	6365.9
99.399	99.416	99.433	99.449	99.458	99.466	99.474	99.482	99.491	99.499
27.229	27.052	26.872	26.690	26.598	26.505	26.411	26.316	26.221	26.125
14.546	14.374	14.198	14.020	13.929	13.838	13.745	13.652	13.558	13.463
10.051	9.8883	9.7222	9.5526	9.4665	9.3793	9.2912	9.2020	9.1118	9.0204
7.8741	7.7183	7.5590	7.3958	7.3127	7.2285	7.1432	7.0567	6.9690	6.8800
6.6201	6.4691	6.3143	6.1554	6.0743	5.9920	5.9084	5.8236	5.7373	5.6495
5.8143	5.6667	5.5151	5.3591	5.2793	5.1981	5.1156	5.0316	4.9461	4.8588
5.2565	5.1114	4.9621	4.8080	4.7290	4.6486	4.5666	4.4831	4.3978	4.3105
4.8491	4.7059	4.5581	4.4054	4.3269	4.2469	4.1653	4.0819	3.9965	3.9090
4.5393	4.3974	4.2509	4.0990	4.0209	3.9411	3.8596	3.7761	3.6904	3.6024
4.2961	4.1553	4.0096	3.8584	3.7805	3.7008	3.6192	3.5355	3.4494	3.3608
4.1003	3.9603	3.8154	3.6646	3.5868	3.5070	3.4253	3.3413	3.2548	3.1654
3.9394	3.8001	3.6557	3.5052	3.4274	3.3476	3.2656	3.1813	3.0942	3.0040
3.8049	3.6662	3.5222	3.3719	3.2940	3.2141	3.1319	3.0471	2.9595	2.8684
3.6909	3.5527	3.4089	3.2587	3.1808	3.1007	3.0182	2.9330	2.8447	2.7528
3.5931	3.4552	3.3117	3.1615	3.0835	3.0032	2.9205	2.8348	2.7459	2.6530
3.5082	3.3706	3.2273	3.0771	2.9990	2.9185	2.8354	2.7493	2.6597	2.5660
3.4338	3.2965	3.1533	3.0031	2.9249	2.8442	2.7608	2.6742	2.5839	2.4893
3.3682	3.2311	3.0880	2.9377	2.8594	2.7785	2.6947	2.6077	2.5168	2.4212
3.3098	3.1730	3.0300	2.8796	2.8010	2.7200	2.6359	2.5484	2.4568	2.3603
3.2576	3.1209	2.9779	2.8274	2.7488	2.6675	2.5831	2.4951	2.4029	2.3055
3.2106	3.0740	2.9311	2.7805	2.7017	2.6202	2.5355	2.4471	2.3542	2.2558
3.1681	3.0316	2.8887	2.7380	2.6591	2.5773	2.4923	2.4035	2.3100	2.2107
3.1294	2.9931	2.8502	2.6993	2.6203	2.5383	2.4530	2.3637	2.2696	2.1694
3.0941	2.9578	2.8150	2.6640	2.5848	2.5026	2.4170	2.3273	2.2325	2.1315
3.0618	2.9256	2.7827	2.6316	2.5522	2.4699	2.3840	2.2938	2.1985	2.0965
3.0320	2.8959	2.7530	2.6017	2.5223	2.4397	2.3535	2.2629	2.1670	2.0642
3.0045	2.8685	2.7256	2.5742	2.4946	2.4118	2.3253	2.2344	2.1379	2.0342
2.9791	2.8431	2.7002	2.5487	2.4689	2.3860	2.2992	2.2079	2.1108	2.0062
2.8005	2.6648	2.5216	2.3689	2.2880	2.2034	2.1142	2.0194	1.9172	1.8047
2.6318	2.4961	2.3523	2.1978	2.1154	2.0285	1.9360	1.8363	1.7263	1.6006
2.4721	2.3363	2.1915	2.0346	1.9500	1.8600	1.7628	1.6557	1.5330	1.3805
2.3209	2.1847	2.0385	1.8783	1.7908	1.6964	1.5923	1.4730	1.3246	1.0000

Appendix 11: Tukey test

$(p=0.05)$

a = the total number of means being compared

v = degrees of freedom of denominator of F test

v \ a	2	3	4	5	6	7	8	9	10
1	17.97	26.98	32.82	37.08	40.41	43.12	45.40	47.36	49.07
2	6.08	8.33	9.80	10.88	11.74	12.44	13.03	13.54	13.99
3	4.50	5.91	6.82	7.50	8.04	8.48	8.85	9.18	9.46
4	3.93	5.04	5.76	6.29	6.71	7.05	7.35	7.60	7.83
5	3.64	4.60	5.22	5.67	6.03	6.33	6.58	6.80	6.99
6	3.46	4.34	4.90	5.30	5.63	5.90	6.12	6.32	6.49
7	3.34	4.16	4.68	5.06	5.36	5.61	5.82	6.00	6.16
8	3.26	4.04	4.53	4.89	5.17	5.40	5.60	5.77	5.92
9	3.20	3.95	4.41	4.76	5.02	5.24	5.43	5.59	5.74
10	3.15	3.88	4.33	4.65	4.91	5.12	5.30	5.46	5.60
11	3.11	3.82	4.26	4.57	4.82	5.03	5.20	5.35	5.49
12	3.08	3.77	4.20	4.51	4.75	4.95	5.12	5.27	5.39
13	3.06	3.73	4.15	4.45	4.69	4.88	5.05	5.19	5.32
14	3.03	3.70	4.11	4.41	4.64	4.83	4.99	5.13	5.25
15	3.01	3.67	4.08	4.37	4.59	4.78	4.94	5.08	5.20
16	3.00	3.65	4.05	4.33	4.56	4.74	4.90	5.03	5.15
17	2.98	3.63	4.02	4.30	4.52	4.70	4.86	4.99	5.11
18	2.97	3.61	4.00	4.28	4.49	4.67	4.82	4.96	5.07
19	2.96	3.59	3.98	4.25	4.47	4.65	4.79	4.92	5.04
20	2.95	3.58	3.96	4.23	4.45	4.62	4.77	4.90	5.01
24	2.92	3.53	3.90	4.17	4.37	4.54	4.68	4.81	4.92
30	2.89	3.49	3.85	4.10	4.30	4.46	4.60	4.72	4.82
40	2.86	3.44	3.79	4.04	4.23	4.39	4.52	4.63	4.73
60	2.83	3.40	3.74	3.98	4.16	4.31	4.44	4.55	4.65
120	2.80	3.36	3.68	3.92	4.10	4.24	4.36	4.47	4.56
∞	2.77	3.31	3.63	3.86	4.03	4.17	4.29	4.39	4.47

v \ a	11	12	13	14	15	16	17	18	19	20
1	50.59	51.96	53.20	54.33	55.36	56.32	57.22	58.04	58.83	59.56
2	14.39	14.75	15.08	15.38	15.65	15.91	16.14	16.37	16.57	16.77
3	9.72	9.95	10.15	10.35	10.52	10.69	10.84	10.98	11.11	11.24
4	8.03	8.21	8.37	8.52	8.66	8.79	8.91	9.03	9.13	9.23
5	7.17	7.32	7.47	7.60	7.72	7.83	7.93	8.03	8.12	8.21
6	6.65	6.79	6.92	7.03	7.14	7.24	7.34	7.43	7.51	7.59
7	6.30	6.43	6.55	6.66	6.76	6.85	6.94	7.02	7.10	7.17
8	6.05	6.18	6.29	6.39	6.48	6.57	6.65	6.73	6.80	6.87
9	5.87	5.98	6.09	6.19	6.28	6.36	6.44	6.51	6.58	6.64
10	5.72	5.83	5.93	6.03	6.11	6.19	6.27	6.34	6.40	6.47
11	5.61	5.71	5.81	5.90	5.98	6.06	6.13	6.20	6.27	6.33
12	5.51	5.61	5.71	5.80	5.88	5.95	6.02	6.09	6.15	6.21
13	5.43	5.53	5.63	5.71	5.79	5.86	5.93	5.99	6.05	6.11
14	5.36	5.46	5.55	5.64	5.71	5.79	5.85	5.91	5.97	6.03
15	5.31	5.40	5.49	5.57	5.65	5.72	5.78	5.85	5.90	5.96
16	5.26	5.35	5.44	5.52	5.59	5.66	5.73	5.79	5.84	5.90
17	5.21	5.31	5.39	5.47	5.54	5.61	5.67	5.73	5.79	5.84
18	5.17	5.27	5.35	5.43	5.50	5.57	5.63	5.69	5.74	5.79
19	5.14	5.23	5.31	5.39	5.46	5.53	5.59	5.65	5.70	5.75
20	5.11	5.20	5.28	5.36	5.43	5.49	5.55	5.61	5.66	5.71
24	5.01	5.10	5.18	5.25	5.32	5.38	5.44	5.49	5.55	5.59
30	4.92	5.00	5.08	5.15	5.21	5.27	5.33	5.38	5.43	5.47
40	4.82	4.90	4.98	5.04	5.11	5.16	5.22	5.27	5.31	5.36
60	4.73	4.81	4.88	4.94	5.00	5.06	5.11	5.15	5.20	5.24
120	4.64	4.71	4.78	4.84	4.90	4.95	5.00	5.04	5.09	5.13
∞	4.55	4.62	4.68	4.74	4.80	4.85	4.89	4.93	4.97	5.01

Appendix 12

The following symbols are the ones we have adopted for use. Whilst most of them are in general use, variations are to be found in statistical literature.

$<$	less than: $3 < 4$.
$>$	more than: $2 > 1$.
x	the numerical value of an observation: e.g. wing length x mm; also the value of a frequency class.
f	frequency; the number of observations in a frequency class x.
y	the numerical value of an observation of a second variable from the same sampling unit from which x is taken: e.g. wing length x mm, tail length y mm.
x^2	x squared.
$\sqrt{x}$	square root of x.
N	number of sampling units in a population.
n	number of sampling units in a sample.
n_i	the 'i'th' observation in a series of n observations.
Σ	capital sigma: the sum of.
Σx	the sum of all values of x in a series of n observations.
$(\Sigma x)^2$	the square of the sum of x in a series of n observations.
Σx^2	the sum of the squares of x in a series of n observations.
μ	mu: population mean.
$\bar{x}$	x bar: sample mean of n values of x.
$\bar{x}'$	x bar primed: derived mean obtained from transformed values of x.
$\bar{y}$	y bar: sample mean of n values of y,
σ	sigma: population standard deviation.
s	estimate of σ from sample data.
σ^2	sigma squared: population variance.
s^2	estimate of σ^2 from sample data.
P	probability.
z	standard deviation unit of the normal curve; test statistic in the z test.
v	nu: degrees of freedom.
H_0	Null Hypothesis.
H_1	alternative to a Null Hypothesis.
χ^2	chi square: test statistic of the chi square test.
r	product moment correlation coefficient of a sample.
ρ	rho: product moment correlation coefficient of a population.
r^2	coefficient of determination.
r_s	Spearman rank correlation coefficient.
a	intercept of a regression line on the y-axis; number of samples being compared.
b	gradient of a regression line (also known as the regression coefficient).
t	test statistic of the t test.
U	test statistic of the Mann-Whitney U-test.
T	Test statistic of the Wilcoxon's test for matched pairs.
F	test statistic of the F test.
q	test statistic of the Tukey test.
k	a parameter of the binomial and negative binomial distributions.
λ	lambda: the parameter of the Poisson distribution.

BIBLIOGRAPHY AND FURTHER READING

Elliott, J.M. (1977) *Some Methods for the Statistical Analysis of Benthic Invertebrates* (2nd edn). Freshwater Biological Association. Publication no. 25.

Moroney, M.J. (1956) *Facts from Figures*. Pelican.

Pentz, M. and Schott, M. (1988) *Handling Experimental Data*. Open University Press.

Pielou, E.C (1975) *Ecological Diversity*. Wiley, New York.

Schott, M. (1990) *Foundation Mathematics for Non-Mathematicians*. Open University Press.

Sokal, R.R. and Rohlf, F. (1981) *Biometry* (2nd edn). Freeman.

Southwood, T.R.E. (1978) *Ecological Methods (2nd edn). Methuen.*

INDEX